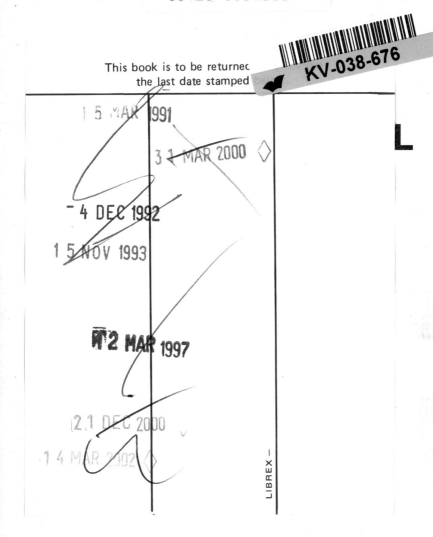

ORGANISATION FOR ECONOMIC CO-OPERATION AND DEVELOPMENT

The Organisation for Economic Co-operation and Development (OECD) was set up under a Convention signed in Paris on 14th December 1960, which provides that the OECD shall promote policies designed:
- to achieve the highest sustainable economic growth and employment and a rising standard of living in Member countries, while maintaining financial stability, and thus to contribute to the development of the world economy;
- to contribute to sound economic expansion in Member as well as non-member countries in the process of economic development;
- to contribute to the expansion of world trade on a multilateral, non-discriminatory basis in accordance with international obligations.

The Members of OECD are Australia, Austria, Belgium, Canada, Denmark, Finland, France, the Federal Republic of Germany, Greece, Iceland, Ireland, Italy, Japan, Luxembourg, the Netherlands, New Zealand, Norway, Portugal, Spain, Sweden, Switzerland, Turkey, the United Kingdom and the United States.

Publié en français sous le titre
SMOG PHOTOCHIMIQUE

© OECD, 1982
Application for permission to reproduce or translate all or part of this publication should be made to:
Director of Information, OECD
2, rue André-Pascal, 75775 PARIS CEDEX 16, France.

CONTENTS

Preface 5

Chapter I
    SUMMARY AND CONCLUSIONS 7

Chapter II
    NATIONAL EMISSION INVENTORIES OF VOLATILE
    ORGANIC COMPOUNDS 12

Chapter III
    VOLATILE ORGANICS EMISSION SOURCES 16

Chapter IV
    AUTOMOBILES: EMISSIONS AND CONTROLS 29

Chapter V
    OIL REFINERIES: EMISSIONS AND CONTROLS 52

Chapter VI
    GASOLINE DISTRIBUTION AND MARKETING:
    EMISSIONS AND CONTROLS 57

Chapter VII
    EMISSION LEVELS FROM OTHER HYDROCARBON SOURCES 66

References 72

Annex
    EMISSION ESTIMATES AND FORECASTS 74

Also available

PHOTOCHEMICAL OXIDANTS AND THEIR PRECURSORS IN THE ATMOSPHERE (September 1979)
(97 78 09 1) 120 pages £3.60 US$7.50 F30.00

THE COST AND EFFECTIVENESS OF AUTOMOTIVE EXHAUST EMISSION CONTROL REGULATIONS (September 1979)
(97 79 04 1) 94 pages £3.20 US$6.50 F26.00

THE COSTS AND BENEFITS OF SULPHUR OXIDE CONTROL (March 1981)
(97 81 01 1) 164 pages £4.80 US$12.00 F48.00

Prices charged at the OECD Publications Office.

THE OECD CATALOGUE OF PUBLICATIONS and supplements will be sent free of charge on request addressed either to OECD Publications Office, 2, rue André-Pascal, 75775 PARIS CEDEX 16, or to the OECD Sales Agent in your country.

PREFACE

Over the past two decades photochemical oxidant pollution has been recognised as a significant environmental problem in a number of OECD Member countries. Photochemical 'smog' is composed of a number of toxic compounds, including ozone, nitrogen dioxide and small particles, which are often referred to as oxidants. They are called secondary pollutants because they are formed in the atmosphere as a result of reactions between certain organic compounds such as hydrocarbons and nitrogen oxides. These two types of substance are the primary pollutants or 'precursors' to the components of photochemical smog.

Most of the reactions between the precursors take place during the warmer months of the year in the presence of direct sunlight. The meteorological conditions conducive to this type of pollution can occur in most parts of the world, and especially between latitudes 60° North and South. Hence, most OECD countries can expect to experience elevated oxidant levels at some time of the year. The problem is of international significance because of the possibility of transport of pollutants across national boundaries and because of the impact of control measures on energy utilisation and the automobile industry.

The main effects of photochemical oxidant pollution are eye irritation, impairment of lung function in people with respiratory diseases, vegetation damage, reduced visibility and various effects on materials, for example, colour fading and attack on rubber products.

The first part of the examination of this problem by the OECD Environment Committee was completed in 1973. This stage took the form of an investigation of the situation in three pilot countries, Australia, Japan and the United States, in which the problem was sufficiently serious to require concerted action in order to avoid human annoyance or health risks, damage to vegetation and effects on certain materials of economic importance (1).

The second stage of the investigation developed from the recognition that the problem was not just peculiar to these three countries but could potentially exist throughout the world since precursor emissions are on the increase, in particular in regions of dense traffic and industry. At this stage (1976) a great deal of information was collected, particularly from Canada and Western Europe and the problem was firmly established as being of global importance. These preliminary investigations established the relative importance of the occurrence of nitrogen oxides, hydrocarbons and related substances in the formation of 'photochemical smog' but no systematic and integrated investigation had been carried out.

It was therefore decided to undertake further work to evaluate the abatement techniques, and their cost, for nitrogen oxides and hydrocarbons; to more fully assess the role of these precursors in the formation of photochemical air pollution; and to examine the type of control strategy required to avoid the resulting problems. This work involved the experience of the OECD Air Management Policy Group, related knowledge obtained from other international organisations and information exchanged during international scientific meetings, and it has already lead to the publication in 1979 of a report entitled "Photochemical Oxidants and their Precursors in the Atmosphere - effects, formation, transport and abatement" (2). A study of the control techniques applicable to nitrogen oxides in stationary sources, their effectiveness and their costs, was completed in 1980.

The present report has been prepared by Mr. J.L. Hedon for the OECD Air Management Policy Group. It identifies major emission sources of volatile organic compounds* and assesses available control methods and their costs for the gasoline cycle, which is responsible for about 50 per cent of such emissions in most Member countries. The report also provides a methodology for estimating volatile organic emissions on the basis of readily available statistics, and presents emission estimates for all OECD countries for 1977, 1985 and 1990.

---

* These particular precursors of photochemical oxidants have traditionally been referred to as 'hydrocarbons' or 'non-methane hydrocarbons' but current preference is for the term 'volatile organic compounds' because not all organic precursors of photochemical oxidants are, in fact, hydrocarbons. Within this report, the terms 'volatile organic compounds' and 'hydrocarbons' are used interchangeably.

Chapter I

SUMMARY AND CONCLUSIONS

DEFINITION AND SCOPE OF THE STUDY

For the purpose of this report, the definition of volatile organic compounds (VOC) given by the US Environmental Protection Agency seems to be the most adequate:

> "A volatile organic compound (VOC) is any organic compound that, when released to the atmosphere, can remain long enough to participate in photochemical reactions. While there is no clear line of demarcation between volatile and non-volatile organics, the predominant fraction of the VOC burden are compounds which evaporate rapidly at ambient temperatures. Almost all organics which can be considered VOC have vapour pressures greater than 0.1 mm of Hg at standard conditions (20°C and 760 mm Hg)."

Furthermore, current preference is to exclude methane, where possible, from VOC emission estimates since it does not participate in photochemical reactions.

Organic compounds are emitted to the atmosphere from both natural and anthropogenic sources. From available data on methane and terpene emissions, it is estimated that worldwide natural emissions are about 2 billion tons per year. There is evidence, however, that many other organic volatiles are also naturally emitted into the atmosphere.

Even though natural organic emissions are greater than anthropogenic emissions on a worldwide basis, natural emissions are not thought to be an important element of the oxidant problem, since they are widely distributed over the entire globe and hence do not add significantly to anthropogenic concentrations in urban or industrialised areas where photochemical oxidant levels can be high. The role of natural emissions in rural areas is not wholly understood because often there is atmospheric transport of organic compounds from urban to rural areas. There is no reason to believe that these natural emissions should vary much in future.

The findings of the 1979 report (2) on 'Photochemical Oxidants and their Precursors in the Atmosphere" were used to estimate the magnitude of the reduction which would appear necessary for anthropogenic emissions. This report considered the US Environmental Protection Agency ozone isopleth method to illustrate the control requirements for different hypothetical areas described by their hydrocarbon/nitrogen oxide ratio (HC/NO_x ratio). Table 1.1 (extracted from the 1979 Report) shows the results obtained when the method is used to calculate the hydrocarbon or nitrogen oxide reduction required in order to reduce to various levels an ozone level of 0.25 ppm, this being the near maximum for a number of cities other than Los Angeles. This table indicates how reduction of one of the precursors (primary pollutants, HC or NO_x) will bring about a certain percentage reduction in the ozone level depending on the ratio of one precursor to the other.

Table 1.1

HYDROCARBON AND/OR NITROGEN OXIDE CONTROL REQUIREMENTS
TO REDUCE 0.25 ppm OZONE TO 0.16, 0.12 OR
0.08 ppm AS A FUNCTION OF HC/NO_x RATIO

Reduce 0.25 ppm ozone to:	Using HC Control only	Using NO_x Control only	Using both HC and NO_x controls
	HC/NO_x = 5		
0.16	43 %	71 %	No advantage(a)
0.12	51 %	88 %	No advantage(a)
0.08	66 %	93 %	No advantage(a)
	HC/NO_x = 10		
0.16	63 %	63 %	50 % each
0.12	69 %	83 %	62 % each
0.08	82 %	89 %	80 % each
	HC/NO_x = 20		
0.16	73 %	57 %	49 % each
0.12	80 %	79 %	67 % each
0.08	90 %	85 %	81 % each
	HC/NO_x = 30		
0.16	78 %	54 %	Little advantage(b)
0.12	84%	74%	Little advantage(b)
0.08	92%	82%	Little advantage(b)

a) Over unilateral HC control.
b) Over unilateral NO_x control.
Source: extracted from (2).

Since the World Health Organisation recommended guideline for photochemical oxidants (measured as ozone) is an hourly average of 0.05 to 0.10 ppm, and a number of countries have already adopted air quality objectives for ozone in this range, the figures in the table for 0.08 ppm are the most relevant. The following significant conclusion can thus be drawn: it is only when the HC/NO_x ratio is below 10 that less than 80 per cent control of HC emissions is required to fulfill the air quality objective of 0.08 ppm ozone, whereas at any HC/NO_x ratio the necessary reduction of NO_x emissions would be greater than 80 per cent.

Following these observations, several questions can be posed:

1. Is it technically possible to obtain such reductions in NO_x and HCs?
2. What is the cost of obtaining such reductions?
3. How will the situation change in future?

This report aims at generating the data base necessary to answer these questions from the point of view of VOC emissions. Reductions in NO_x emissions are considered in a separate report mentioned earlier.

EMISSION INVENTORIES

Before reviewing available control techniques and their costs, and assessing the control requirements necessary to achieve the desired air quality objectives, the magnitude of VOC emissions in OECD countries must be determined.

During the early 1970s, a number of countries produced national estimates of VOC emissions from anthropogenic sources (see Chapter II). These national emission inventories were compiled for different years using different methodologies and source categories. Therefore, from the data available, it was not always possible to identify which sources were included or to obtain a comparable breakdown of emissions by source category and product or fuel type. Hence, it was not possible to produce future emission forecasts on the basis of these inventories, nor even to consider the basic estimates as comparable in any sense.

To make emission forecasts feasible, as well as to obtain estimates for all OECD countries, a methodology has been developed using only readily available statistics which would enable: (a) definition of source categories, and (b) determination of emission factors for these categories and, where possible, their likely evolution over time.

This methodology is described in Chapters III to VII and the resulting emission estimates are given in the Annex. The energy data used to calculate emissions are taken from the OECD publication "Energy Balances of OECD Countries" (24) as well as from OECD Secretariat energy forecasts for 1985 and 1990. Given the variety of VOC sources, it was impossible to consider every possible source or national specification. Generalisations and 'rule of thumb" relationships were developed when data were not available so that a complete picture of VOC emissions could be obtained for all OECD countries. The limitations of this exercise are well recognised and the emission levels presented in the Annex should be considered more as orders of magnitude than as exact figures.

CONTROL TECHNOLOGY

The most significant VOC emission category, 'the gasoline cycle', which includes refinery emissions, gasoline distribution and marketing, and motor vehicle emissions, represents 50 per cent or more of total VOC emissions in most OECD countries. The gasoline cycle is also a sector for which control technology already exists on a large scale. VOC emission reduction measures in the oil industry may imply significant economic and energy savings, and so they have often been introduced without legal requirement. In addition, there is legislation related to motor vehicle emissions in most OECD countries which requires the introduction of control technology.

The report includes a detailed discussion of control technologies and costs for each emission source in the gasoline cycle. This cycle is here divided into: motor vehicle emissions - Chapter IV; oil refinery emissions - Chapter V; and gasoline distribution and marketing - Chapter VI. Emission factors for 1985 and 1990 reflect the extent to which controls, or additional controls, are expected to be introduced by these dates.

Other VOC emission sources are generally small individual emitters for which the introduction of controls often raises technical and economic problems. Emission factors for these sources are discussed in Chapter VII, but a detailed discussion of control technologies and costs is not included because: (i) the principles of these technologies, based on incineration, adsorption, absorption or condensation, are well known; (ii) actual emission reductions and cost calculations usually require a specific case-by-case approach, which would not have been possible within the framework of this project; and (iii) emission levels from these sources are not expected to vary considerably on a national scale up to 1990 because of the difficulty and high cost of stringent control for most such sources.

CONCLUSIONS

1. It is estimated that natural emissions of volatile organic compounds (VOC) are one order of magnitude greater than anthropogenic emissions on a worldwide basis. However, their contribution to

the photochemical smog pollution problem is not believed to be significant because natural emissions are widely distributed over the entire globe and are largely composed of methane (at least 70 per cent) which is usually non-reactive photochemically.
2. Major anthropogenic VOC sources, and their relative importance, have been identified. The two largest emitters are the gasoline cycle (including motor vehicle fuel processing and distribution) and solvent evaporation. For the former, emission sources have been identified and quantified and possible control technology is discussed; for the latter, the magnitude of emissions has only been estimated.
3. A simple methodolgy based on available energy statistics, population data and 'Gross Domestic Product' growth rates is presented in this report to enable countries to estimate their VOC emissions. For a more precise inventory on a regional or local scale, modifications can be introduced to reflect local conditions.
4. Estimates of VOC emissions have been made for all OECD countries and three different years (1977, 1985 and 1990) on a sector-by-sector basis and by fuel type (see Annex).
5. The available VOC control technologies are already used without legal obligation on a number of sources where product recovery provides an economic incentive. However, because VOC emissions originate mostly from fossil fuels and their derivatives, they may increase drastically in future if no additional control action is taken.

Chapter II

NATIONAL EMISSION INVENTORIES OF VOLATILE ORGANIC COMPOUNDS

During the early 1970s, a number of countries produced national estimates of VOC emissions from anthropogenic sources. It is to be noted that, even though in many of these countries there were no government regulations limiting VOC emissions from stationary sources, some industries (for example the petroleum industry) were already applying control technologies because of the value of recovered organic products. For mobile sources, many countries imposed emission control regulations in the 1970s, but initially these had little effect because they only applied to a small proportion of the vehicle population.

Table 2.1 summarises these national emission estimates. More detailed information was available for three countries:

- For Japan, Automobile Emission estimates for Tokyo and Tokyo Bay Area for the years 1965 to 1990 are presented in Table 2.2.
- For Sweden, details of the transport sector VOC emissions for the year 1975 are given in Table 2.3.
- For the United States, each emissions source category is broken down into its component parts and details for the year 1972 with projections to 1985 and 1990 are given in Table 2.4.

Table 2.1

NATIONAL EMISSIONS OF VOLATILE ORGANIC COMPOUNDS

in 10^3 Metric Tons

Country/year	Australia(a) 1972	Canada(b) 1972	Canada(b) 1974	Germany(c) 1971	Germany(c) 1975	Italy(d) 1972	Japan(e) 1973	Netherlands(f) 1975	Netherlands(f) 1985	Norway(g) 1970	Norway(g) 1975	Norway(g) 1980	Swed.(h) 1975	U.K. 1975	United States(i) 1972	United States(i) 1985	United States(i) 1990
1. Transportation	506	1,521	1,320	325	762	647	n.a.	271	327	131	156	186	252	660	12,500	7,000	6,400
2. Fuel Combusion in Stationary sources	33	31	147	246	176	25	14	19	24	2	1	2	13	40	1,500	1,700	2,100
3. Petroleum Refining	16	133	133				198							160	0,900	0,700	0,700
Natural Gas Processing		10	11														
(Other) Industrial Processes	62	3	4	955	869	538	76	490	584	4	3	6	60	120	8,400	12,500	15,600
4. Gasoline Marketing	101	188	158										8				
5. Degreasing							85						30				
Dry Cleaning		29	30				117						5				
Surface Treatment		190	217				825						64				
Evaporation of Solvents	186									Negl.				310	4,300	4,600	5,200
6. Solid Waste Incineration	203	46	30							55	64	78	35		1,100	600	500
7. Miscellaneous (forest fires, open-burning, etc.)	111	412	449												900	900	900
Total	1,218	2,563	2,499	1,526	1,807	1,210	1,315	780	935	192	224	272	432	1,325	29,600	28,000	31,400

a) Australia - National Report, Department of the Environment and Conservation, Canberra City, February 1974.
b) Canada - 1972: National response to OECD questionnaire, 1977
   1974: National contribution to 1978 OECD project.
c) Federal Republic of Germany -
   1971: Second Report on the Problem of Photochemical Oxidants and their Precursors in the Atmosphere, OECD, Paris, 1976.
   1975: Preliminary data from response to OECD questionnaire.
   It is not known if VOC emissions from solid waste incineration and burning, gasoline storage and marketing, and forest and structural fires are included in the total. It is assumed, however, that "Industrial Processes" includes all evaporation of solvents from surface coating, dry cleaning and metal degreasing.
   The decrease in VOC emissions from fuel combustion in stationary sources cannot be easily explained since fuel consumption increased by about 9 per cent from 1971 to 1975 but with slight shift in the percentage share from coal to gas.
d) Italy - Second Report on the Problem of Photochemical Oxidants and their Precursors in the Atmosphere, OECD, Paris, 1976.
e) Japan - Japanese Environment Agency (see also Table 2.2).
f) Netherlands - National Report on Oxidants in the Netherlands, September 1975.
   The 1985 data are based on national estimates.
g) Norway - Emissions of Nitrogen Oxides and Hydrocarbons in Norway, Bjørg Fjeld, Norwegian Institute for Air Research (NILU), Norway, June 1974.
   Miscellaneous includes forest fires, open burning, dry cleaning, surface coating, and gasoline distribution and marketing estimated as 30-50 per cent of total of other categories (from reference).
h) Sweden - National response to OECD questionnaire, 1977 (see also Table 2.3).
i) United States - US Environmental Protection Agency (Data extracted from Table 2.4).

Table 2.2

VOC EMISSION ESTIMATES FROM
MOTOR VEHICLES IN TOKYO AND TOKYO BAY AREA
as of March 1977

10^3 metric tons

Year	Tokyo	Tokyo Bay Area
1965	87.8	151.8
1966	98.2	172.6
1967	111.6	201.3
1968	124.7	229.8
1969	139.7	263.5
1970	153.4	300.0
1971	142.9	299.0
1972	133.4	280.6
1973	126.7	270.1
1974	104.5	231.9
1975	95.2	214.7
1976	85.5	196.3
1977	77.4	179.5
1978	71.1	165.1
1979	66.4	153.2
1980	62.9	143.5
1985	58.5	127.6
1990	62.8	137.1

N.B.: 1. Figures after 1975 are based on estimates.
2. Future estimates of emission factors are based on future regulations, i.e. the regulations adopted in 1973 and 1975.

Table 2.3

VOC EMISSIONS FROM THE TRANSPORT SECTOR IN SWEDEN, 1975

metric tons

Passenger cars, buses, trucks, motorcycles, mopeds and motor boats:	
Passenger cars, gasoline	127 300
Passenger cars, diesel	1 650
Buses, gasoline	1 850
Buses, diesel	1 600
Trucks, gasoline	12 375
Trucks, diesel	9 650
Motorcycles	3 000
Mopeds	15 750
Motor boats, gasoline	51 850
Motor boats, diesel	120
Motor tools, tractors and transport trucks used in the following branches:	
Agriculture, gasoline engines	1 900
Agriculture, diesel engines	2 375
Forestry, gasoline engines	15 350
Forestry, diesel engines	875
Fishing, gasoline engines	200
Fishing, diesel engines	500
Industry, gasoline	4 125
Industry, diesel	1 575
Total	252 045

Table 2.4

UNITED STATES NATIONWIDE VOC EMISSIONS WITH SELECTED CONTROLS[a]

10^6 metric tons/year

Source Category	1972	1985	1990
Transportation	12.5	7.0	6.4
Highway vehicles	10.9	5.0	4.1
Non-highway vehicles	1.6	2.0	2.3
Stationary fuel combustion	1.5	1.7	2.1
Electric utilities	0.1	0.1	0.1
Industrial	1.3	1.5	1.9
Residential, commercial and institutional	0.1	0.1	0.1
Industrial processes	9.3	13.2	16.3
Chemicals	2.2	3.7	5.0
Petroleum refining	0.9	0.7	0.7
Metals	0.2	0.2	0.2
Mineral products	0 0	0.1	0.1
Oil and gas production and marketing	2.9	3.9	4.5
Industrial organic solvent use	2.9	4.4	5.6
Other processes	0.2	0.2	0.2
Solid waste	1.1	0.6	0.5
Miscellaneous	5.2	5.5	6.1
Forest wildfires and managed burning	0.7	0.7	0.7
Agricultural burning	0.2	0.2	0.2
Coal refuse burning	0 0	0 0	0 0
Structural fires	0	0	0
Miscellaneous organic solvent use	4.3	4.6	5.2
Total	29.6	28.0	31.4

N.B.: A zero indicates emissions of less than 50,000 metric tons per year

a) Developed by the U.S. Environment Protection Agency. 1972 emissions are projected to 1985 and 1990 assuming compliance with current regulations and future emission targets that have already been legislated for (i.e. Federal Motor Vehicle Control Program, and New Source Performance Standards as of 1977).

Chapter III

VOLATILE ORGANICS EMISSION SOURCES

INTRODUCTION

Volatile organic compounds (VOC) are emitted from a wide variety of sources. The relative importance of each of these VOC sources for oxidant formation depends on the amount and location of the emissions, and sometimes on the specific type of VOC emitted: this last aspect is important because of the relative reactivity of a compound and its subsequent role in photochemical reaction processes.

The major VOC emission sources are grouped into six categories: motor vehicles, solvent evaporation, oil industry, incineration, manufacture of chemicals, and combustion at stationary sources.

I. MOTOR VEHICLES

Motor vehicles are the primary source of VOC emissions in all OECD countries, accounting for 40 to 60 per cent of the total anthrogenic emissions.

The gasoline powered internal combustion engine is an inherently high emission system. This is largely because of the fuel and the difficulty of bringing about its complete combustion.

a) <u>The Gasoline Engine</u>

Emissions from a gasoline fuelled vehicle are of three types, namely evaporative, crankcase and exhaust, with more than half of the total emissions coming from the exhaust.

Evaporative VOC emissions come from the fuel tank and the carburettor system. Diurnal changes in ambient temperature result in expansion of the air-fuel mixture in a partly-filled fuel tank and, as a result, gasoline vapour is expelled to the atmosphere. Running losses from the fuel tank occur as a result of reflected heat from the road surface, and hot-soak losses from the carburettor system occur after the engine stops. Carburettor losses are from, for example, the carburettor vents, the float bowl, and around the throttle and choke shafts. Crankcase emissions come from numerous points in the crankcase ventilation or lubrication systems of the engine.

An analysis of exhaust VOC emissions begins with the air/fuel mixture coming from the carburettor. This mixture is ignited in a combustion chamber by a spark. The various characteristics of the combustion process, such as temperature, duration etc. depend upon air/fuel ratio, engine design and operating conditions.

The burnt mixture of air and fuel consists mostly of carbon dioxide (CO_2) and water vapour (H_2O), but because of incomplete combustion, carbon monoxide (CO) is formed and some VOCs remain uncombusted. In addition, some of the fuel can "blow-by" adding to the VOC emissions, and nitrogen oxides are also formed by the reaction of nitrogen (N_2) and oxygen (O_2) in air which is encouraged by the high temperature in the combustion chamber.

The actual amount of VOC emissions depends on factors such as air/fuel ratio, spark timing and quality, combustion chamber design, engine speed and load and engine temperature.

b) The Diesel Engine

In the diesel engine, the fuel is ignited by compression of the gases instead of by the spark used in the gasoline engine. Most VOC emissions are unburnt gases resulting from incomplete combustion. This occurs because the temperature in the cylinders is too low for fuel ignition or because of quenching on the walls of the cylinders.

The VOC emissions from diesel engines are less than from gasoline engines, but diesel engines also emit a black or blue smoke and an unpleasant odour if they are operated at overload, poorly designed, tuned or maintained.

II. SOLVENT EVAPORATION

a) Dry Cleaning (3)

Dry cleaning is the cleaning of fabrics using non-aqueous organic solvents. The process requires three steps: (i) washing the fabric in solvent, (ii) spinning to extract excess solvent, and (iii) drying by tumbling in a hot airstream. The solvent itself produces the VOC emissions.

Two general types of cleaning fluids are used in the industry: petroleum solvents and synthetic solvents. Petroleum solvents are inexpensive, combustible hydrocarbon mixtures similar to kerosene. Synthetic solvents are non-combustible but more expensive halogenated hydrocarbons, generally Perchloroethylene and trichlorotrifluorethane. It is generally economically desirable to recover and reuse the synthetic solvents; recovered petroleum solvents may be reused or combusted.

There are two basic types of dry cleaning machines - transfer and dry-to-dry. Transfer machines accomplish washing and spinning in one unit and drying separately. Dry-to-dry machines are single units that perform all three operations. All petroleum solvent machines are the transfer type, but synthetic solvent plants can be either.

b) <u>Surface Coating</u> (3)

Surface-coating operations primarily involve the application of paint, varnish, lacquer, or paint primer for decorative or protective purposes. This is accomplished by brushing, rolling, spraying, flow coating or dipping. Automobile and aircraft assembly, manufacturers of containers, furniture, appliances, and plastic products, job enamelers and automobile repainters, are among the industries involved.

Hydrocarbon emissions occur in surface-coating operations because of evaporation of paint vehicles, thinners, and solvents used to facilitate application of the coatings. The main factor determining these emissions is the volatile content of the coating, most, if not all, of which is evaporated and therefore emitted during application and drying of the coating. The compounds released include aliphatic and aromatic hydrocarbons, alcohols, ketones, esters, alkyl and aryl hydrocarbon solvents and mineral spirits.

c) <u>Paint Manufacture</u> (4)

Paint manufacture involves the following operations: mixing pigments with a vehicle to form a paste, grinding and diluting the paste, tinting to required colour, testing, straining, filling and packaging. Paint manufacturing is still largely a batch process because of the great variety of raw materials and finished products required. Many of these products must be custom formulated and processed.

The two operations producing VOC emissions are grinding and thinning. During grinding, heat is produced which causes vaporisation of certain organic ingredients. In the thinning operation, vaporisation of solvent occurs. Thinning of premixed paint pastes to the required consistency involves dilution with aliphatic or aromatic hydrocarbons, alcohols, ketones, esters, and other highly volatile materials. Because of the volatility of most thinners, mixing must be done in totally enclosed tanks to prevent solvent loss.

d) <u>Metal Degreasing</u> (5)

In many industries, metal articles must be washed or degreased before being electroplated, painted, or otherwise surface finished.

Solvent degreasers vary in size from simple unheated wash basins to large heated conveyorised units in which articles are washed in a hot solvent vapour spray which is the most common type of industrial degreaser. Vaporisation is carried out by electricity, steam, or gas heat.

Non-flammable chlorinated solvents such as perchloroethylene ($Cl_2C = CCl_2$) and 1,1,1-trichloroethane (CH_3CCl_3) or trichloroethylene ($ClHC = CCl_2$) are almost always used for degreasing. Selection of solvent is usually dictated by the temperature necessary to dissolve grease and tar.

III. OIL INDUSTRY

a) <u>Production of Crude Oil and Gas</u> (4)

The production and processing of oil for transport to refineries involves recovery of fluids from the well, gas-oil separation, oil-water (brine) separation and storage.

Crude oil in the production field is generally stored in both bolted and welded steel tanks, usually vertical with a fixed roof. Floating roofs are seldom used in the production field.

Major sources of VOC emissions include evaporation from brine pits and tanks, incomplete flaring, and leaks. For every kiloliter of oil extracted, an additional two to three kiloliters of waste brine are produced. Waste brine may contain some residual oil and usually has a concentration of dissolved solids seven times that of seawater. The most common method of disposal is reinjection, although it may be dumped in rivers, marine waters, unlined pits, non-potable water sites or approved disposal sites. The open disposal methods allow free evaporation of hydrocarbons. Waste water separators enable recovery of some hydrocarbons, but they also produce VOC emissions. With offshore production, the water is either cleaned before discharge into the sea or pumped into tankers or pipelines for treatment onshore.

Natural gas is composed of methane with decreasing amounts of ethane, propane, and heavier hydrocarbons. Removal of these heavier components is a necessary step in producing natural gas for pipeline sales. Hydrocarbon emissions from natural gas processing are mainly fugitive in nature and result from leaks in pumps, valves, compressors, and other machinery. Such emissions are about 89 per cent methane.

b) <u>Crude Oil Transfer</u> (4)

Crude oil is transported from production areas to refineries by pipeline, tanker, barge or tank car.

Emissions from pipeline transfer as well as ship and barge loading and ballasting operations will be discussed in this section. Emissions from storage and tank cars and trucks are discussed under gasoline distribution and marketing.

i) <u>Pipelines</u>

The primary sources of VOC emissions during pipeline transfer are from compressor stations and fugitive emissions from pipelines.

Compressor stations are used to maintain flow through long-distance pipelines. They often use natural gas fired internal combustion engines which emit large quantities of unburnt organics. Methane makes up a large proportion of these emissions.

The current trend is towards decreased use of these engines and their replacement by electric motors with greater reliability and lower costs.

Another source of emissions from pipeline transportation is leaks. Under the influence of heat, pressure, vibration, friction and corrosion, leaks can develop in the packing surrounding pipeline valve stems.

ii) <u>Ship and barge terminals</u>

Marine terminals are generally located at the end of pipelines or adjacent to refineries and chemical plants. The four major sources of VOC emissions from marine terminals are storage tanks, leaks, loading operations, and ballasting operations.

Ship and barge loading is the largest source of emissions from marine terminal operations. Loading losses occur as VOC vapors in empty cargo tanks are displaced to the atmosphere when the liquid is introduced.

The cruise history of a cargo carrier is another important factor in loading losses. Emissions are generally lowest when the cargo tanks are free from vapors prior to loading. This occurs when previous cargo was a non-volatile product or because of tank cleaning or ventilation.

Non-segregated ballasting is the second largest source of organic emissions from marine terminals. Cargo tanks on large tankers are often filled with water after cargo is unloaded in order to improve stability of the empty tanker. Emissions occur because organics-laden air in the empty cargo tank is displaced to the atmosphere by ballast water. However, when separate ballast tanks are employed to store ballast water, no such emissions occur.

c) Petroleum Refining Operations

Since oil refineries are constructed for the handling and processing of large volumes of organic compounds, they will inevitably emit some hydrocarbons into the atmosphere. Although a refinery is a complex system of many processes, emissions generally result from processing, effluent water treatment, storage and loading and unloading processes.

i) Processing

The entire processing operation can be divided into four major steps: separating, converting, treating and blending. The crude oil is first separated into selected fractions (e.g. gasoline, kerosene, fuel, oil, etc.). Because the relative volumes of each fraction produced by merely separating the crude may not meet the market demands for each fraction, some of the less valuable products, such as heavy naphtha, are converted into products of greater economic value, such as gasoline. This conversion is accomplished by splitting (cracking), uniting (polymerisation), or rearranging (reforming) the original molecules. The final step is the blending of the refined base stock and incorporation of various additives to meet final product specifications.

Most of these refining processes, including regeneration of any necessary catalysts, form essentially closed systems and thus there are relatively few major point sources of VOC emissions. However, all refining processes involve a number of small potential sources, of which the most important ones are described below:

Pressure relief valves:

Process equipment is normally protected against overpressure by relief valves. These spring-loaded pressure relief devices may blow down into the flare system or be connected to a vapour control device. Where these are not vented to an enclosed system, leakage to the atmosphere can occur because of corrosion or incorrect setting.

Pumps and Compressors:

These can leak product at the point of contact between the moving shaft and the stationary casing.

Valves and Flanges:

As already stated in the section on pipeline transfer, pipeline valves can leak at valve stems, and flanges can leak because of incorrect installation of the gasket material.

Refinery Flares:

A flare is used to burn hydrocarbon vapours which are vented to prevent unsafe operating pressures or during unit start-ups and shutdowns. Under normal refinery operating conditions, only a small amount of gas is burnt in the flare. This combustion is almost complete and thus only small amounts of hydrocarbons are emitted. In modern flares, steam injection is used to assist combustion and reduce smoke formation. Thus, significant emissions occur only during abnormal conditions when sufficient steam may not be available. Such conditions are comparatively infrequent. The multiflare system is another common technique used to avoid VOC release. In this system, the normal (small) quantity of flare gas is burnt in one flare but when the pressure rises, more flares are commissioned.

Catalytic Cracking Units:

Catalytic cracking uses heat, pressure, and a catalyst to convert heavy oils into lighter products. Catalytic cracking processes currently in use can be classified as either fluidized bed (FCC) or moving bed (TCC) units. Organic emissions from catalytic cracking operations result from catalyst regeneration.

Other Sources:

A major source of emissions from crude oil fractionation processes in the past was the barometric condenser on vacuum distillation columns which allowed non-condensable hydrocarbons to escape into the atmosphere. Modern vacuum installations are equipped with shell and tube type condensing equipment and are no longer a significant contributor to atmospheric process emissions.

ii) Effluent Water Treatment

Oil contaminated effluents (tank bottoms, ballast water and seal coolant, etc.) are generated at all stages of refinery operation. Primary treatment of oily effluents is practised in all refineries. This takes the form of gravity separators or interceptors designed to allow the oil to rise to the surface of the aqueous phase for removal by skimming or floating-off. Most refineries are equipped with the American Petroleum Institute (API) type separator. Hydrocarbons can escape to the atmosphere from the large, open area of these separators. Generally, these oil-water separators are major emitters of organics.

iii) Storage

Two types of tanks are commonly used for the storage or liquid hydrocarbons with a true vapor pressure (TVP) not exceeding 800 mbar (above this TVP pressure, storage tanks are normally used). The two types are termed fixed roof and floating roof.

Fixed roof tanks are satisfactory for products of low volatility such as fuel oils and diesel fuels. However, when more volatile liquids, such as crude oils and gasoline, are loaded into fixed roof tanks hydrocarbon gases in the vapor space are expelled to the atmosphere through the tank vent; these losses are termed working losses. In addition, there are breathing losses from the vapor space due to thermal expansion of the tank contents as ambient temperature fluctuates.

In floating roof tanks, the roof remains in contact with the liquid surface at all times and vapor losses are about 20 times lower than from a fixed roof tank on a similar duty with volatile products. These losses, termed standing storage loss, occur through the gap between the floating roof and the tank shell. A small loss results from the evaporation of liquid adhering to the tank shell when the roof is lowered during product withdrawal.

Floating roof tanks are used on volatile product duty to minimise atmospheric emissions, to conserve hydrocarbons, and for safety reasons.

There are, at present, technological limits to the retrofitting of floating roofs or covers on fixed roof tanks. It is difficult to maintain a good seal and easy operation on new tanks under 4 to 5 meters diameter because of lack of stability for the roof or cover. For old fixed roof tanks with a small capacity, the original design and/or irregular walls may prevent retrofitting. In Japan, however, it is intended to equip all tanks of 1,000 kiloliter capacity or more with floating roofs.

Finally, it should be noted that floating roofs are not a requirement in many OECD countries, but they are in general use because of their economical benefits in terms of reduction of product or crude loss. Consequently, the future trend will be towards an increasing proportion of floating roof tanks.

d) Gasoline Distribution and Marketing

In marketing terminals, the main losses occur during gasoline storage and handling; emissions from other products are very slight. Because of the relatively small size of individual sites compared to refineries, certain gasoline tanks are of the fixed roof rather than floating roof type.

In marketing operations, the main losses occur in the gasoline distribution chain. The emissions consist of vapors actually present in the vessel at the start of loading, together with vapors generated from, or by, the material being loaded; they are generally termed "displacement" emissions and tend to arise during certain operations, namely loading at terminals, discharging at filling stations and vehicle refuelling at filling stations.

In Europe, transit losses have been shown to be very low during normal operation.

IV. INCINERATION

a) <u>Incineration of Industrial, Commercial and Municipal Refuse</u>

According to the U.S. Solid Waste Disposal Act of 1965, "solid waste" may be defined as garbage, refuse and other discarded solid materials resulting from industrial, commercial and agricultural operations, and from community activities. In large urban areas, for health reasons and space constraints, solid waste incineration has long been an economical disposal option.

It is difficult to estimate the total amount of waste produced in OECD countries. For the U.S., one estimate states that the per capita generation rate of urban and industrial waste is approximately 4.5 kg/day*, half of which is burned. This figure may vary from country to country depending on consumption habits, extent of industrialisation, etc.

In most instances, incineration is not a complete waste disposal method in itself since there is an ash or residue which remains after combustion. After incineration, however, the volume of solid waste is reduced considerably and can be handled easily by conventional land disposal methods.

Incinerator type and size can vary greatly:

Municipal incinerators are often multiple-chamber units equipped with automatic loading mechanisms, temperature controls and movable grate systems. VOC emissions are generally fairly low.

Industrial and commercial incinerators are of either single or multiple-chamber design; these units are often manually loaded and intermittently operated. Some industrial incinerators are similar to municipal incinerators in size and design, but most are smaller. Better designed emission control systems include gas-fired afterburners or scrubbing, or both. Emissions are generally higher than for municipal incinerators because of design parameters and lower combustion temperature.

b) <u>Open Burning</u> (2)

Open burning may be done in open drums or baskets, in fields and yards, and in large open dumps or pits. Materials commonly disposed of in this way are municipal waste, auto body components, landscape refuse, agricultural field refuse, wood refuse, bulky industrial refuse, and leaves.

Ground-level open burning is affected by many variables including wind, ambient temperature, composition and moisture content of the debris burned, and compactness of the pile. In general, the relatively low temperatures associated with open burning increase VOC emission.

---
* Survey conducted by the U.S. Public Health Service.

Emissions from agricultural refuse burning are dependent mainly on the moisture content of the refuse, how much refuse material is burned per unit of land area and how the refuse is arranged.

Of the total wood refuse from lumber mills and wood-working industries, only a small proportion can be processed into useful products such as chip board, paper, etc.; the bulk of this waste is disposed of by incineration, open burning or landfilling.

c) <u>Forest Wildfires</u> (2)

The size and intensity (or even the occurrence) of a wildfire are directly dependent on such variables as local meteorological conditions, tree species and their moisture content, and the weight of consumable fuel per acre (fuel loading). Once a fire begins, the dry combustible material (usually small undergrowth and forest floor litter) is consumed first, and if the energy release is large and of sufficient duration, the drying of green, live material occurs with subsequent burning of this material as well as the larger dry material. Under suitable environmental and fuel conditions, this process may initiate a chain reaction that results in a widespread conflagration.

V. MANUFACTURE OF CHEMICALS

The main branches of the chemicals industry include the manufacture of plastics, paints and varnishes, soap and detergents, pharmaceutical products, dyestuffs, sulphuric acid and feedstocks like ethylene and propylene. Most processes have potentially high emissions, but because of the economic value of the compounds the emissions are usually recovered and recycled. Some other processes use closed systems allowing little or no emission to the atmosphere.

In a number of OECD countries, the petrochemical industry accounts for a significant share of total emissions from chemical manufacturing processes. Given the complex and ever-changing nature of this industry, it is not possible to include in this report an exhaustive list of VOC emitting processes and equipment; in addition, only limited information is available on VOC emissions from the chemical industry in general.

Those branches of the chemicals industry which are considered important in Japan, the United Kingdom and the U.S. in terms of emissions are given in the following tables:

Table 3-1

VOC EMISSIONS FROM SOME CHEMICAL
MANUFACTURING PROCESSES IN JAPAN (1973)

10^3 metric tons

Manufacturing process	Annual Production		Annual emissions(b)
	Nationwide	P.I.A.(a) affiliated companies (%)	
Ethylene	4,170	4,034 (90)	0.460 (0.7)
Polyethylene by high pressure	1,011	1,011 (100)	3.950 (5.7)
Polyethylene by mid pressure	612	506 (83)	2.540 (3.6)
Styrene monomers	935	858 (91)	0.990 (1.4)
Ethylene oxide	405	341 (85)	2.150 (3.1)
Acetic aldehyde	611	313 (51)	0.460 (0.7)
Acrylonitrile	598	595 (100)	46.400 (66.5)
Polypropylene	688	688 (100)	8.880 (12.9)
Synthesized rubber	967	784 (81)	3.940 (5.6)
Benzene, toluene, xylene	3,972	2,355 (59)	0.030 (0.0)
Total	13,964	11,465 (82)	69.800 (100)

a) P.I.A. = Petrochemical Industry Association.
b) Annual emissions are those from P.I.A. affiliated companies and not nationwide emissions.

Source: P.I.A. survey.

Table 3-2

ESTIMATES OF VOC EMISSIONS FROM SOME CHEMICAL
PROCESSES IN THE UNITED KINGDOM (1972)

Manufacturing process	Annual production (10^3 tons)	Annual emissions (10^3 tons)	Type of emission
Synthetic rubber	300	11	Butadiene, styrene, acrylonitrile
Plastics			
- PVC	300	1	Vinyl chloride, solvents
- Polypropylene	200	0.03	Propylene, solvents
Synthetic fibres			
- Nylon	100	0.1	Solvent
Phthalic anhydride	60	0.9	Phthalic and maleic anhydrides, quinone
Paint (1974)	440,000	44	Solvent
Total		57.03	

Source: Reference (6)

Table 3-3

VOC EMISSIONS FROM CHEMICAL INDUSTRIES
IN THE UNITED STATES (1977)

Source	Estimated emissions(a) 10^3 metric tonnes
Industrial Processes	
Paint manufacture	24
Vegetable oil processing	15
Pharmaceutical manufacture	50
Rubber products manufacture	140
Plastic products manufacture	No estimates available
Textile polymers manufacture	No estimates available
Others	750
Organic Chemical Manufacture	
Process streams	450
Storage and handling	300
Waste disposal	150
Fugitive (leaks)	600
Major Chemical Products(b)	
Acrylonitrile	69
Ethylene oxide	42
Ethylene dichloride	34
Dimethyl terephthalate	33
Vinyl chloride	18
Ethylene	16
Propylene oxide	16
Ethylbenzene	16
Methyl methacrylate	15
Formaldehyde	12
Methanol	12
Maleic anhydride	11
Terephthalic acid	9
Acetic acid	9
Cyclohexane	7
Acetic anhydride	7
Vinyl acetate	6
Carbon tetrachloride	6
Butadiene	5
Phenol	5
Acetone	5
Cyclohexanol/Cyclohexanone	5
Chloroprene	5
Ethylene glycol	5
Acrylic acid	5

a) EPA estimates for 1977.
b) EPA estimates. These numbers include only process stream emissions and in-process storage and handling emissions. Waste disposal, fugitive, and out-of process storage and handling emissions are not included.

Source: Reference (4).

VI. COMBUSTION IN STATIONARY SOURCES (4)

Stationary fuel combustion sources may utilise external or internal combustion. External combustion sources include boilers for steam generation, heaters for the heating of process streams, and driers and kilns for the curing of products. Internal combustion sources include gas turbines and reciprocating internal combustion engines.

External combustion sources are categorised according to the type of fuel burned in the unit; coal, fuel oil and natural gas are the main fuels used.

VOC emissions from stationary external combustion sources are dependent on type and size of equipment, method of firing, maintenance practices, and on the grade and composition of the fuel. Considerable variation in organic emissions can occur, depending on the efficiency of the individual unit; incomplete combustion results in more emissions.

Internal combustion engines include gas turbines or large heavy-duty reciprocating engines. Most stationary internal combustion engines are used to generate electric power, to pump gas or other liquids, or to compress air for pneumatic machinery. The primary fuels are natural gas and distillate fuel oil, although residual oil is sometimes used. Emissions from gas turbines are considerably lower than emissions from reciprocating engines; however, reciprocating engines are generally more efficient.

The organic emissions from stationary internal combustion sources result from incomplete combustion of the fuel. Emissions from compression engines, particularly reciprocating ones, are significantly greater than those from external combustion boilers.

Chapter IV

AUTOMOBILES: EMISSIONS AND CONTROLS

In most developed nations, in the 1960s or early 1970s, the significance of automobiles emissions (hydrocarbons HC, nitrogen oxides NO_x, and carbon monoxide CO), particularly in urban areas, was realised. To control these emissions, the most effective instrument has been emission standards. A review of the existing legislation in OECD countries indicates that there are three main sets of automobile emission standards for the US, EEC, and Japan. Various other countries have also adopted legislation which generally follows, with a few years' delay, US legislation or EEC Directives.

The selection of a standard has a direct influence on the type of control technology introduced for automobiles and on costs incurred. As standards become more stringent, the control technology becomes more complex, and costs increase. Since automobiles are very large contributors to total VOC emissions, the selection by countries of a particular control programme for such emissions will be a determining factor in future VOC emissions.

Before presenting in detail the automobile emission control options, it should be pointed out that emission standards are aimed at reducing not only HC emissions but those of NO_x and CO as well. Most control techniques and engine designs raise the problem of the trade-off between reducing one pollutant and increasing another; this is one of the main difficulties in vehicle emissions control since reducing emissions of all three pollutants requires opposing strategies. Consequently, emission controls on automobiles have been developed to achieve a balance in the reduction of these pollutants, and technologies and costs for one of these pollutants (in this case HC) cannot be easily separated.

Control options for automobiles can be divided into: control technologies (or primary measures) affecting the design of vehicles; and secondary measures which include inspection/maintenance programmes and traffic reduction.

A. CONTROL TECHNOLOGIES

1. The gasoline engine

In gasoline-fuelled automobiles, there are three major sources of HC emissions: combustion exhaust, crankcase blowby, and evaporative loss from the fuel system. Table 4.1 illustrates the relative contributions of each, as estimated by the U.S. EPA (7), for uncontrolled vehicles of the 1963-1967 model years.

Table 4.1

PASSENGER CAR HYDROCARBON EMISSIONS

	Exhaust		Blowby		Evaporative	
	g/mile	g/km	g/mile	g/km	g/mile	g/km
Pre-control (a)	9.1	5.7	4.1	2.5	2.53	1.57
Percentage of total	57.9		26.1		16.0	

a) Low altitude, 49-state passenger cars.

Crankcase emissions are those from any part of the crankcase ventilation or lubrication systems. These have essentially been eliminated with "Positive Crankcase Ventilation Systems". In these systems, air is circulated through the crankcase and the air and blowby gases are drawn into the intake manifold, where they are carried to the combustion chambers. They perform well when maintained, but can totally fail if the critical control valve becomes clogged.

Evaporative loss originates from the fuel tank and carburettor. These emissions can be reduced with a vapor recovery system connected to the crankcase or an adsorption-regulation system which condenses and returns vapors to the fuel tank, or traps them until they can be fed back into the engine induction system.

Many techniques (8,9) are now being used to attain low exhaust emissions. These include methods of achieving almost complete combustion, either by altering the fuel supply to the engine or the combustion process itself, and techniques which act on the exhaust gases once they have left the engine.

Modification of combustion chamber design: A major source of HC emissions from automobiles is unreacted fuel-air mixtures expelled through the exhaust. This occurs primarily because the thin layer of gaseous mixture which makes contact with the relatively cool combustion chamber surfaces does not burn. Almost complete combustion of the full cylinder charge is promoted by modifying combustion chamber design to reduce the surface-to-volume ratio and by minimising crevices.

Modification of induction system: CO and HCs in the exhaust often result from insufficient oxygen in the air/fuel mixture and consequent incomplete combustion. Leaner air/fuel mixtures to assure more complete combustion can be achieved by converting more of the liquid gasoline into vapour form, and by enabling improved air/fuel mixing and distribution among the cylinders. Air/fuel induction systems can be adapted to provide heated intake air for more uniform carburettor inlet temperatures, thus allowing the use of leaner air/fuel mixtures. More uniform distribution of the air/fuel mixture to the cylinders can be accomplished through intake manifold heating or through design changes. Unfortunately, modifications of induction systems which improve combustion and reduce HC and CO emissions also raise temperatures and contribute to higher NO_x emissions.

Carburettor modifications: The air/fuel ratio directly affects the output, fuel economy, and smooth operation of an engine; it also considerably affects the concentration of exhaust emissions. The carburettor is therefore a key element in effective emission control because of its role in metering the fuel in proper proportion to inlet air. Precise fuel metering, in accordance with changing engine requirements, enables operation with leaner air/fuel mixtures. Carburettors can be designed with better fuel metering and reduced calibration tolerances to ensure better fuel mixing.

Choke modifications: Gasoline in liquid form does not burn. Therefore, when an engine is started cold, extra gasoline is required to obtain enough vapourised HCs to mix with air and provide a combustible mixture at the spark plug. The carburettor choke supplies the extra fuel. Unvapourised HCs, however, pass through the engine unburned. By tailoring choke action to car requirements, enrichment during starting and warm-up can be made compatible with satisfactory performance over a wide temperature range.

Ignition system modification: Correctly modified ignition systems bring improvements in fuel metering and mixture control. Spark retardation can be used to reduce HC and NO_x emissions. Electronic ignition systems have been developed which improve control of spark timing in all operating conditions, facilitate spark timing adjustment on vehicles in use, and have greater reliability. Retarding ignition timing results in more fuel being burned during the exhaust phase of the combustion cycle. Accordingly, some loss in power and fuel economy results, and demands on the engine cooling system are increased.

Lower compression ratio: The use of high compression ratios improves engine efficiency and results in greater power output for a given amount of fuel. Combustion temperatures are high, however, causing high NO_x emissions. The octane requirements of high compression ratio engines are high, necessitating lead addition or fuel modification. The presence of lead in gasoline severely limits the effectiveness of catalytic converters and reduces the life of other emission control system components.

Exhaust gas recirculation (EGR): Recirculation of a portion of the exhaust gas into the air/fuel mixture reduces the formation of NO_x. Dilution of the fuel charge with inert gases has the secondary effect of reducing engine octane requirements, but with some loss in power. Excess dilution causes misfiring and poorer performance. Improved systems enable proportioning of recirculated exhaust gas to the air flow demanded by the engine.

Fast burning: This system combines several engine modifications to promote low HC, NO_x and CO emissions, while maintaining good fuel economy. While high EGR is used to decrease NO_x emissions, strong turbulence in the air/fuel mixture and two spark plugs on each cylinder enable maintenance of stable combustion, low HC and CO emissions, and good performance. However, it does require modification to the intake manifold, ignition system, and combustion chamber design.

Electronically controlled fuel injection system: These are computer-controlled devices which inject the correct amount of gasoline for the amount of intake air into each cylinder so as to maintain the appropriate air/fuel ratio.

Stratified charge engine: This gasoline-fuelled internal combustion engine differs from the conventional engine mainly in combustion chamber design, and the use of fuel injection. In this system, a precombustion chamber is located within the combustion chamber, with lean mixture supplied to the main chamber and rich mixture supplied to the precombustion chamber. The flame generated by the spark ignition in the prechamber burns the mixture in the main chamber. The aim is to reduce NO_x formation through the combustion of a lean mixture; CO and HC are controlled through an oxidation reaction by the use of a heat-retaining manifold. Exhaust emissions could conceivably be low because combustion occurs with excess air. There could be a problem, however, in maintaining effective stratification at any load; if the most remote and lean parts of the charge refuse to burn, unburned fuel in the exhaust can be excessive, resulting in reduced thermal efficiency.

Air injection: After the gases have left the combustion chamber, air is injected as close as possible to the exhaust valve where any CO and HCs present are hot enough to ignite immediately, and their oxidation is thus completed. Changes of the cylinder head and exhaust manifold are required. Since control of NO_x during the combustion process has tended to increase HC and CO emissions, interest in exhaust port air injection is reviving. Air injection pumps are also helpful for effective operation of thermal reactors and catalytic converters.

<u>Thermal reactors</u>: A thermal reactor functions as a combustion chamber outside the engine, and normally appears in the form of an oversized exhaust manifold. Thermal reactors receive the hot exhaust gas from the engine and, by insulation, retain as much heat as possible. Additional heat is generated by oxidation of CO in the exhaust gases. With these devices, the engine can be operated with rich fuel mixtures producing high CO concentrations. Such reactors are known as "rich thermal reactors". Supplementary air is often required, and appropriate mixing and adequate residence time are essential for the reaction of the combustibles present with oxygen. When designed for rich air/fuel mixtures to promote NO_x control, there is a substantial fuel penalty. In a "lean thermal reactor" system, the carburettor is set lean so that the exhaust is inherently oxidising, and a secondary air pump is not required; however, emissions are generally higher than from "rich" reactors.

Because extremely high temperatures can be reached, materials of suitable durability must be used. Special protective systems are needed to prevent engine damage from overheating.

<u>Catalytic converters</u>: Catalytic converters are devices designed to convert exhaust gases into harmless substances by chemical oxidation or reduction. The catalyst bed generally consists of an active material deposited in a thin layer on an inert support. The emission control systems presently being used in the U.S. and Japan generally use oxidation catalysts to reduce CO and HC emissions. The next generation of system will employ either a reduction catalyst system as the first stage, or a three-way catalyst system in order to reduce NO_x emissions as well. The reduction catalyst system is based on combustion of a rich mixture, with NO_x controlled through a catalytic reduction reaction, while CO and HC are controlled through catalytic oxidation with secondary air supply. NO_x are generally further decreased by EGR. In the three-way catalyst system, the three pollutants (CO, HC, and NO_x) are controlled simultaneously by one catalyst, with NO_x further controlled by EGR. Three-way catalysts, however, require accurate control of stoichiometric air/fuel ratio. This is accomplished by precise fuel metering in conjunction with several sensors including an oxygen sensor in the exhaust and an electronic control unit. These recent systems may reduce catalyst deterioration caused by engine malfunction or maladjustment. Like all catalytic converters, however, they require unleaded gasoline; catalyst poisoning by use of leaded fuels remains a major concern.

2. The diesel engine

CO and HC emissions from diesel engines are much less than from gasoline engines, although NO_x emissions are the same. Diesel engines emit black or blue smoke if they are poorly designed, tuned or maintained, and exhaust gases may have an unpleasant odour. Diesel smoke, however, can be reduced substantially by using additives, by proper engine maintenance, avoidance of overloading, by engine derating, and adherence to fuel specifications.

The following improvements on conventional engines are used to control CO, HC, NO_x and diesel smoke: (i) in the intake and exhaust systems, modification of intake manifold shape and improvement in intake port and valve timing; (ii) in the fuel supply system, improvement in the fuel injection system and better injection timing; (iii) improvement in combustion chamber geometry.

The effect of these control measures varies for each pollutant. CO, HC, and diesel smoke are reduced with improved combustion, while NO_x decrease must be achieved by lowering combustion temperature to control oxidation. These conflicting requirements must somehow be met to achieve overall emission control.

Since the fuel in the diesel engine is burned by self-ignition, its combustion is more difficult to control than in the gasoline engine. Furthermore, a large air intake results in greater variation of the air/fuel ratio. It is also necessary to take the smoke limit into consideration for the determination of maximum output. These factors make it extremely difficult to control CO, HC, and NO_x simultaneously in the diesel engine.

3. Control technology costs (11)

The calculation of automobile emission control costs is very complex because of the many parameters which influence the average cost. The costs can be divided into three groups:

- initial cost;
- lifetime maintenance cost (10 years/160,000 km);
- fuel consumption cost.

Initial cost

Because the automobile industry reacts to other legislation (safety and fuel economy) as well as environmental regulations, it is difficult to estimate the direct initial cost of emission control. For instance, although electronic engine control plays a part in emission control, it is also installed to improve safety and fuel economy. Thus, all its costs should not be attributed to a specific programme. As a result, one can create a wide range of cost estimates simply by juggling the cost percentage applicable to a particular programme. This is probably the largest source of error in estimations carried out by government and industry with the latter tending to overestimate, and the former to underestimate control costs to support their own vested interests and goals. Moreover, mobile source emission control systems are designed to reduce three pollutants (i.e., HC, CO and NO_x). Consequently, it may not be appropriate to assign the total cost of motor vehicle controls solely to HC control. In the recent cost analysis associated with the review of the ozone standard, the U.S. EPA assumed that only one-third of the cost of mobile source controls was attributable to HC control.

Estimates developed in various U.S. studies in the 1972-74 and 1976-77 periods are compared in Table 4.2. It can be seen that the initial estimates in 1972-73 generally overestimated the eventual costs. It was not until a 1974 report by the National Academy of Sciences (NAS) (11) was published that a more accurate method of estimation was developed. This improved accuracy stems from a sophisticated estimation model which "assembles" a vehicle, and costs the process according to unit costs sensitive to production volume, hardware, configuration, material costs and vehicle size.

The NAS costs are based on a 6-cylinder intermediate car which represents the "average" sized car in the U.S., and thus approximates the average cost. The NAS costs compare very well with many of the latest cost estimates made in 1976-77.

The 1979 U.S. emission standard was the same as for the 1977: 0.92 g/km HC, 9.3 g/km CO, and 1.2 g/km NO_x. The estimated uncontrolled new vehicle emission levels existing in the 1960s were: 5.4 g/km HC, 54 g/km CO, and 2.5 g/km NO_x. With the objective of controlling emissions of photochemical oxidant precursors, the U.S., in 1979, was using technology which reduced new vehicle HC and NO_x emissions by 83 per cent, and 52 per cent respectively. The cost estimates for this reduction are given in the 5th line of Table 4.2.

According to a recent U.S. EPA study (12), the predominant emission control system expected to be in production for the 1980s (beginning with the 1981 emission standards of 0.25 g/km HC, 2.1 g/km CO, 0.6 g/km NO_x) is a three-way plus oxidation catalyst system using feedback carburation, exhaust gas recirculation and air injection. Some vehicles, particularly those of inertia weight of 3,000 pounds (1,360 kg) or less and with low power-to-weight ratios, could be certified using current oxidation catalyst technology. In this study, the initial cost to the consumer of these systems is given as $158-$162 (13) for a vehicle with an oxidation catalyst system; and $269-$285 (13) for a vehicle with a three-way plus oxidation system.

Control costs must be considered in terms of the effectiveness of emission reduction over vehicle lifetime and not just in relation

Table 4.2

U.S. ESTIMATES OF CUMULATIVE INITIAL COSTS OF EMISSION CONTROLS(1)

Emission standard (g/km)			Estimated proto-type control level (g/km)			Estimates made in 1972-74				Estimates made in 1976-77			
HC	CO	NOx	HC	CO	NOx	EPA(a)	NAS(b)	Industry(c)	NAS(d)	EPA(e) (1976)	Industry(e)	General Motors(f)	EPA(g)(2) (1977)
3.7	31.6	(3.1)	1.1	9.5	(2.2)	8	9						
1.9	17.4	(3.1)	0.57	5.2	(2.2)	49	55						
1.9	17.4	1.9	0.57	5.2	1.3	116	76	116	58	33– 98	140		
0.93	9.3	1.9	0.28	2.8	0.84	249	223	221–284	120	33–129	140–318	213–250	33–129
0.93	9.3	1.2	0.28	2.8	0.84					33–129	174–275		
0.56	5.6	1.2	0.17	1.7	0.84				178	72–145			
0.25	5.6	1.2	0.08	1.7	0.43								
0.25	5.6	0.62	0.08	1.7	0.43	375	314	326–438	209	180–221	163–339	238–275	68–239
0.25	2.1	1.2	0.08	0.63	0.84					180–225	291–511	303–360	178–319
0.25	2.1	0.62	0.08	0.63	0.43							353–410	68–254
0.25	2.1	0.25	0.08	0.63	0.17	459	492	305–600	360	285–330	176–637	353–410	233–389
													285–469

1) All costs in 1977 US dollars.
2) The EPA study used the 1977 standards as baseline and then calculated incremental costs (i.e. 0.93, 9.3, 1.2 as baseline). To adjust the estimates to uncontrolled cars, the 1976 EPA estimates from reference e were used as the baseline values.

Sources:

a) U.S. EPA, *The Economics of Clean Air*, report to U.S. Congress, Document no. 92-67, Washington, 1972.
b) National Academy of Sciences, *Report by the Committee for Motor Vehicle Emissions*, Washington, 1973 (N.B. Estimate made in 1972).
c) Battelle Memorial Institute, *Cost of Clean Air 1974*, NTIS PB-238 762, Springfield, Va., 1974 (N.B. Estimate made in 1972).
d) National Academy of Sciences, *Report by the Committee on Motor Vehicle Emissions*, Washington, 1974.
e) U.S. EPA, *Automobile Emission Control – The Current Status and Development Trends as of March 1976*, Washington, 1976.
f) General Motors Corporation, *Published Estimated Costs*, 1977.
g) U.S. EPA, *An Analysis of Alternative Motor Vehicle Emission Standards*, Washington, 1976. (N.B. These costs are sales weighted and the difference represents a cost versus fuel optimised system in 1980.)

to new vehicle standards. The general conclusion of a review of various emission control surveys in the U.S., Canada, Japan, Sweden and the EEC was that there is a rapid and serious reduction in effectiveness of control over vehicle lifetime.

Using a five-year-old car as the indicator of the average lifetime emission rate, and the factors developed by the U.S. EPA to represent emission rates by level of control and age, Table 4.3 has been developed (10). It shows that the cars currently in operation are functioning with average emissions over twice the original new vehicle standards.

Therefore, because of degradation of emission control equipment over vehicle lifetime, the actual reduction in emission by the control technology in current use in the U.S. is much lower. Instead of 83 per cent, HCs have actually been reduced by 68 per cent from uncontrolled levels. For NO_x, the actual reduction is only 10 per cent from the uncontrolled levels in in-use vehicles of 2.1 g/km instead of the 55 per cent which represents the reduction from the uncontrolled new vehicle level of 2.5 g/km. These lower values represent the impact of degradation in the fifth year of operation of the vehicle.

Table 4.3

EPA ESTIMATED AVERAGE IN-USE EMISSION RATES[a]
FOR CURRENT CONTROL SYSTEMS

Original standard (g/km)			In-use emission rate (g/km)			In-use rate/standard		
HC	CO	NO_x	HC	CO	NO_x	HC	CO	NO_x
	uncontrolled		5.0	54	2.1	–	–	–
3.7	31.6	–	3.4	40	2.9	0.92	1.26	–
1.9	17.4	–	3.4	40	2.9	1.79	2.30	–
1.9	17.4	–	3.4	40	2.1	1.79	2.30	1.11
0.93	9.3	1.9	1.6	22	2.0	1.72	2.37	1.05
0.93	9.3	1.2	1.6	22	1.9	1.72	2.37	1.58

a) Average emission rates in fifth year of operation as measured by CHS-CH method.

Source: Reference (3).

Another cost data source was the study done by the German Government (14) to support its proposed 1982 emission standards. These proposals call for a joint HC+NO_x standard of 10 grams per test (ECE test procedure), while the CO standard ranged from 7.5 to 12 g/km (in terms of the U.S. test procedure) based on inertia weight. The grouping of the HC and NO_x into one standard allows the manufacturer more design flexibility. The ratio of HC to NO_x should not be expected to change substantially from the 1/3 to 2/3 ratio currently attained. Assuming this ratio is sustained, then these proposed German standards can be translated into effective design standards of 0.65 g/km HC and 1.0 g/km NO_x (in terms of the U.S. test procedure). The estimated cost of attaining these standards was set at US $160 by the UBA. If these standards are related to the U.S. prototype control levels (see Table 4.2), it is apparent that they are comparable - as far as CO and HC emissions are concerned - with the U.S. 1975-77 standards (lines 4 and 5 of Table 4.2), and - as far as NO_x emissions are concerned - with the 1973-75 U.S. standards (lines 3 and 4). The UBA estimates are thus very similar to the U.S. estimates. It should, however, be mentioned that the UBA estimates are not limited to the reduction of exhaust emissions, but include efforts to reduce fuel consumption at the same time.

Initial cost estimates have also been made by the Japanese Government (15). They reported that the price increase between motor vehicles complying with 1973 standards and those complying with 1975-76 standards was about 10 per cent. This means an average price increase per car of approximately $250, assuming a 300 yen/dollar exchange rate for that time period. Although this price rise involves the variables of raw materials and labour costs and profit margins, it gives us a useful basis for estimating the costs of automotive exhaust controls and their economic impacts. Although there has been no comparison of the 1973 or 1975-76 Japanese standards, there has been a comparison of the 1978 Japanese standards with the U.S. EPA test procedure (16). Recognising the difficulties of comparing standards based on two different procedures, the report concluded that the 1978 Japanese standards were equivalent to 9.0/0.9/0.6 (CO/HC/NO_x) g/km on the U.S. test. With this in mind, it is suspected that the 1975-76 Japanese standards were closely comparable to the 1977 U.S. EPA standard. The figure of $250 can therefore be compared with cost estimates on line 5 of Table 4.2.

Lifetime Maintenance Cost

There have been relatively few studies of maintenance cost increases associated with emission controls. In general, they have attempted to estimate the cost increase due to the maintenance for emission controls. As with initial price, the maintenance items which are included can vary depending on the analysts' perspective; for example, spark plug replacement could be included even though this was a requirement before the advent of emission controls shown in Table 4.4. The data have all been adjusted to reflect the estimated per-vehicle costs over 10 years or 160,000 km.

The costs given in Table 4.4 do indicate some consensus among governments about maintenance costs.

Since many of the studies calculate the cost under the assumption that all required control system maintenance is done, the cost

Table 4.4

ESTIMATES OF LIFETIME CONTROL MAINTENANCE COSTS

Emission standard (g/km)			Effective design standard (g/km)			Estimated lifetime control maintenance costs (1977 US $)		
HC	CO	NO_x	HC	CO	NO_x	EPA(a)	NAS(b)	UBA(c)
3.7	31.6	(3.1)	1.1	9.5	(2.2)	224	0	
1.9	17.4	(3.1)	0.57	5.2	(2.2)	224	327	90-320
1.9	17.4	1.9	0.57	5.2	1.3	224	387	
0.93	9.3	1.9	0.28	2.8	1.3	252(1)	255(1)	
0.93	9.3	1.2	0.28	2.8	0.84			
0.56	5.6	1.2	0.17	1.7	0.84		270(1)	
0.25	5.6	1.2	0.08	1.7	0.84			
0.25	5.6	0.62	0.08	1.7	0.43			
0.25	2.1	1.2	0.08	0.63	0.84	280(1)	270(1)	
0.25	2.1	0.62	0.08	0.63	0.43			
0.25	2.1	0.25	0.08	0.63	0.17	392(2)	460(2)	

1) Assumes a catalyst change at 80,000 km costing $135.
2) Assumes a catalyst change at 80,000 km costing $371.

Sources:
a) U.S. EPA, reference (17).
b) National Academy of Sciences, reference (11).
c) Umweltbundesamt, reference (14).

estimates will be higher than what is actually incurred by the consumer. This is particularly true in the EPA and NAS estimates in Table 4.4 where the cost of catalyst remplacement is added to the lifetime costs. It is extremely unlikely that this will occur, since the present regulations in the U.S. and Japan do not mandate such a replacement. It is worth noting that in more recent EPA estimates the cost of catalyst replacement has not been included.

Maintenance costs are very sensitive to the amount of lead in the gasoline. Although the addition of lead is beneficial as far as fuel economy and vehicle performance are concerned, the lead and the chemical scavengers present act as contaminants and corrosive agents in the engine and exhaust system. U.S. studies (Table 4.4) indicate a zero or negative cost with advanced controls, which is largely a result of the lead-free fuel used.

The German data from UBA indicate a good correlation with the U.S. data in the same control regime. They estimate that their newest 1980 standard proposals would increase lifetime maintenance costs by £90–$320.

Table 4.5

MAINTENANCE CHANGES OVER 160,000 km
FROM UNCONTROLLED CAR TO 3-WAY CATALYST SYSTEM

Change O_2 sensor 3 times	3 x 15 = $45
Miscellaneous emission system repairs	$50
Save 5 plug changes	5 x 10 = $(-) 50
Save 10 point/condenser changes	10 x 10 = $(-)100
Save 1 muffler change	$(-) 20
Total	$(-)75 per 10 years or $(-)7.5 per year

Fuel Consumption Cost

The sensitivity of fuel economy to emission control is related to many factors of engine design and control, such as spark timing, air/fuel ratio, EGR type and compression ratio. Because of the range of effects of these factors, knowing the direction in which one is moving in the control of exhaust emissions is not sufficient to deduce the direction change in fuel economy. However, certain emission control devices (catalytic converters and thermal reactors) do not effect fuel economy while reducing emissions and hence the engine can be calibrated for optimal fuel economy, while reduction of exhaust concentrations is left to the after-treatment device. Because these devices exist, EPA has concluded that: "At a fixed emission level, fuel economy is a function of the usage of fuel efficient control technology" (18).

The U.S. has found that there is effectively no fuel economy penalty inherent in the standards up to and including those of 1977. There will, however, be transient fuel economy penalties as the industry learns to optimise vehicle designs for both cost and fuel economy. In fact, the U.S. EPA is currently estimating that there will be little or no fuel economy loss at standard levels as stringent as 0.25 HC, 2.1 CO, 0.62 NO_x (g/km), provided adequate lead time for development and optimisation is allowed.

The general findings by EPA are in conflict with the conclusions of a study undertaken for the ECE by the Bureau Permanent International des Constructeurs d'Automobiles (BPICA)(19). BPICA estimated that there would be a 1.2-3.3 per cent increase in fuel consumption over the range of emission standards it investigated.

Control levels which BPICA investigated are approximately those achieved by the American Programme in 1970-1973 period. Although for North American cars, at that time, there was up to 10 per cent loss in fuel economy, this loss, according to EPA's analysis, was due entirely to the use of non-optimal control systems. BPICA has, in fact, produced estimates representing the cost-optimal vehicle design which would create a penalty in fuel economy associated with standards. Thus, it seems that they have estimated the highest possible impact on fuel consumption.

4. Cost Effectiveness of Control Techniques In-use

The data presented in this chapter provide a limited basis for assessment of cost effectiveness of control techniques in use today.

The 1977 U.S. new automobile exhaust emission standards are estimated to have an effectiveness of 68 per cent reduction for HC emissions, and 10 per cent for NO_x emissions.

The 1973 U.S. standards, which are currently in use in several other countries, are estimated to have an effectiveness of 32 per cent reduction for HC emissions, and zero per cent reduction for NO_x emissions.

The 1975-76 Japanese standards probably have an effectiveness similar to that of the U.S. 1977 standards but, so far, there are only limited data to support this. It should also be determined whether the Japanese practice of a 30,000 km durability requirement plus enforced in-use inspection maintenance is more or less effective in preventing degradation than the U.S. requirement of 80,000 km durability and no in-use inspection.

For Germany, the UBA report showed HC emissions decreasing by 45 per cent and NO_x emissions increasing by 60 per cent from 1970 to 1975 on the basis of 45 vehicles tested under the ECE test procedure and standards. A limited Swedish testing programme on high mileage vehicles indicated a substantial deterioration to levels which could be construed as close to uncontrolled. It can only be concluded that ECE standards are at present less effective compared with U.S. and Japanese standards.

The total cost of emission control (initial and lifetime maintenance) to obtain the emission levels of either the 1977 U.S. standards, the 1975-76 Japanese standards or the proposed 1982 German standards can be estimated at $350 to $450 per average vehicle. This assumes a fuel-efficiency-optimised design for the emission control system and catalyst replacement at 80,000 km. If the system is not fuel-efficiency-optimised but instead cost-optimised and an increase of 1.2-3.3 per cent in fuel consumption is incurred (as predicted by BPICA), then the total cost per vehicle would increase by $110 to $320 in additional fuel consumption, assuming a 160,000 km lifetime. The range is mostly dependent on gasoline price which, in OECD countries, ranges from about $0.30/1 to $0.80/1. However, these would add only to the lower end of the total cost range, i.e. to $350, since the cost would be optimised, and hence, low.

For countries following the 1973 U.S. standards, the total cost per average vehicle can be estimated to be about $300 to $400 with the same assumptions on fuel-efficiency-optimised design. These emission control systems do not use catalytic converters.

In conclusion, for fuel-optimised emission control systems, the cost effectiveness of catalytic converter systems is $5 to $7 per 1 per cent HC reduction, and 0.15 per cent NO_x reduction. For non-catalyst-based systems, (i.e. the U.S. 1973 standards), the cost effectiveness is $9 to $12 per 1 per cent HC reduction with no NO_x reduction. To avoid any misunderstanding of these cost-effectiveness data, it is worth noting that mobile source control costs assume the application of a certain control technology. Cost variability depends on the selection of the technology and not only on the percentage HC reduction. For example, assuming a catalytic system were used, the cost per vehicle would be the same in a country that needed a 30 per cent reduction as in a country that needed a 40 per cent reduction.

B. SECONDARY CONTROL MEASURES

1. Inspection/Maintenance

Motor vehicle inspection and maintenance (I/M) is a programme of periodic inspection to determine emission levels; vehicles found to emit excessive amounts of pollutants are failed, and must be repaired and reinspected.

The cost of building an inspection station varies according to the size of the station, the cost of land, and the type of test run.

For a two-lane, centralised public facility using the "idle test", the capital cost of the station ranges from $117,000 to $237,000. For a "loaded test", where the vehicle is run under simulated driving conditions on a dynamometer, the capital cost is estimated at $140,000 to $260,000 (20).

The maintenance costs per serviced vehicle vary inversely with the percentage of vehicles which fail the test. At a 30 per cent failure rate, the weighted average cost per vehicle is estimated at $8.24 (2). Based on this result and on estimates of current effectiveness of I/M in reducing hydrocarbon emissions, the cost-effectiveness of I/M has been estimated at $635 per ton of hydrocarbon removed (12).

2. Traffic Reduction Plans

These plans generally include various measures aimed at reducing pollution load in congested urban areas, as well as reducing overall fuel consumption by motor vehicles. These measures are basically aimed at discouraging low-occupancy automobile use and at encouraging transit and carpool use.*

Carpools

Carpools are an effective means of reducing commuter vehicle-miles travelled (VMT). With four members in a carpool, VMT can be reduced by 75 per cent of the VMT if each person drove separately.

The individuals can also make sizeable savings in travel costs. For a four-person carpool with a 20-mile round-trip, one report estimates that each rider will make annual savings of $475. This saving increases to $925 for a 40-mile round-trip (21).

The primary costs associated with carpooling are promotional costs. An areawide computerised carpool matching service, however, can be operated for about $4 per participant (21).

Dedicating lanes on motorways and city streets for the exclusive use of buses and carpools during peak travel periods enables these vehicles to bypass congested sections of roadways. This increases the attractiveness of high-occupancy travel modes since travel time is substantially reduced. Costs can vary widely depending on existing infrastructure and on the complexity of necessary improvements.

Public Transport Improvements

A number of aspects of public transport can be improved to enhance the level of service. These include marketing, security measures, shelters, terminals, and fare policies and fare collection

---

* "Cost and Economic Impact Assessment for Alternative Levels of the National Ambient Air Quality Standard for Ozone" (Draft), June, 1978.

techniques. Since these measures are dependent solely on local conditions, no attempt is made to assess costs.

Parking Management

Parking restrictions, when coupled with transport and carpool incentives, may be an effective means to attain and maintain standards. Parking management policies can have a dramatic effect on traffic flow and VMT. These include: (1) the location of parking, (2) the amount of on- and off-street space allocated to parking, (3) the charges applied, and (4) the length of time parking is permitted.

One means to discourage parking is through the use of taxes or surcharges to increase parking cost. Studies in the U.S. have predicted that an increase in daily parking cost up to one dollar can result in a reduction of 3 to 15 per cent in VMT in the central business district, with an accompanying increase in public transport use (21).

Supplementary Licensing

In city centres or other places where traffic congestion is serious, vehicle movements may be diverted and reduced by obliging drivers entering the congested area to purchase and display a special supplementary vehicle licence (22). The need to display such licences may be confined to peak travel times or to other periods. Licence-free passage may be given to buses and bicycles, and to cars carrying more than a certain number of people in order to promote higher occupancy rates and to avoid penalising lower income car owners. Enforcement may be limited to recording the registration numbers of vehicles not displaying licences at points of entry to the controlled area, but alternative solutions are possible.

C. EMISSION FACTORS

Present and future HC emissions from motor vehicles have been estimated from available national data and information on motor vehicle emission control programmes.

The U.S. and EEC control programmes were used as the basis for calculation of the emission factors because: (1) most information is available for these two programmes; (2) they are the most representative in terms of numbers of countries using them; and (3) while a number of OECD countries implement programmes that are different from those mentioned above, the technical and historical data were not sufficient to allow separate calculations. Consequently, only two scenarios have been considered - one for countries with or considering catalyst technology, and one for other countries.

It is recognised that motor vehicle HC emissions of some countries will be somewhat over- or underestimated because of this procedure, but on the other hand, it has the merit of giving an overall picture for the whole of the OECD.

The United States Programme

The following approach has been used to obtain HC emission factors for the transportation sector for the three years considered (1977, 1985, 1990).

The transportation sector has been divided into highway vehicles (cars, lorries, buses, motorcycles) and non-highway vehicles (aircraft, barges and vessels, trains, etc.). This division of the transportation sector requires some comment:

1. Each of the two derived emission factors will represent an average emission factor for different types of vehicle regrouped into one category.
2. These two emission factors will vary each year as a function of the mix of model years (and consequently of controls).
3. Finally, the two emission factors will also reflect (at least for historical data) different degradation factors because of the deterioration of specific controls due to age and mileage of the vehicles.

From available HC emission estimates in the U.S. (23) and corresponding fuel consumption data (24), the emission factors on Table 4.6 were obtained for 1970-1977. The highway vehicle emission factor (EF) decreased from 37.1 kg/Toe in 1970 to 26.8 kg/Toe in 1977 because of the gradual introduction of controls. The non-highway EF varied around 24 kg/Toe (between 25.7 and 23.1); a constant and equivalent decrease cannot be observed for such vehicles, however, since they have not been subject to these controls.

Table 4.6(a)

HC EMISSION FACTORS IN THE UNITED STATES

	1970	1971	1972	1973	1974	1975	1976	1977
Highway	37.1	35.5	34.3	31.9	30.3	28.2	27.9	26.8
Non-highway	25.7	24.5	23.3	23.6	23.1	23.2	24.9	23.7

a) For more details on how these factors were developed, see Appendix to this chapter.

The 1972 highway HC EF in Table 4.6 has been compared with the national average emission factor for highway vehicles given in reference (3) for 1972. Assuming standard conditions as given by EPA (31.6 km/h, 20 per cent cold operation at 24°C), and an average consumption in 1972 of 20 litres/100 km, the EPA factor of 6.7 g/km becomes 35.8 kg/Toe, i.e., a 4 per cent difference from the 1972 figure in Table 4.6.

The 1985 and 1990 emission factors have been developed from the U.S. EPA nationwide emission estimates presented in Chapter II and internal secretariat energy forecasts. The emission factor for non-highway vehicle has been assumed constant (24 kg/Toe), whereas the EF for highway vehicles reflects projected results of the U.S. Federal Motor Vehicle Control Programme (FMVCP). These average emission factors for highway vehicles, then, are 13.7 kg/Toe and 10.0 kg/Toe for 1985 and 1990, respectively; i.e., about 50 per cent and 60 per cent reduction from 1977 HC emissions.

From a U.S. EPA study (16), it appears that present U.S. and Japanese motor vehicle standards are comparable in stringency and from what is known of future goals this comparability will probably continue. A number of OECD countries follow more or less exactly the U.S. control programme with a few years' delay, or have adopted stringent control programmes based on the "catalyst technology".

Because of this similarity in conditions and lack of specific national data, the HC emission factors developed for the U.S. have been applied to all "catalyst technology" countries, even though a country-by-country approach would be more desirable.

The EEC Programme

The EEC motor vehicle emission control programme is significantly different from both the U.S. and the Japanese programmes, and, as such, has led to a separate evaluation for present and future HC emission factors.

As mentioned in Chapter II, recent, comprehensive and comparable emission data do not exist for all European countries. Some national data exist, but they were not calculated on the same basis and are often not sufficiently detailed. It is, therefore, difficult to compare and use these data in the same way as the U.S. data.

To overcome this problem, a German estimate (14) of HC emissions from an average European vehicle has been used to obtain emission factors for highway vehicles equivalent to U.S. vehicles: the basis for this "European" value is an emission factor of 38.4 g HC per kg fuel (41.2 kg/Toe) for an uncontrolled vehicle. Table 4.7 has been developed on the basis of three factors: this figure for uncontrolled vehicles, the implementation of the EEC Directives, and 10 per cent replacement of the car population each year (i.e., a 10-year lifetime).

No specific information was available on degradation of control requirements under the EEC programme. However, it appeared from

Table 4.7(a)

HC EMISSION FACTORS FOR THE EEC

	1970	1971	1972	1973	1974	1975	1976	1977
Highway vehicles	41.2	41.2	40.5	39.8	39.2	38.6	37.6	36.8

a) For more details on how these factors were developed, see Appendix to this chapter.

tests* in the U.S. and in Sweden that the degradation of emission control could vary significantly. For early control levels in the U.S. (up to about 30.35 per cent control), no degradation was observed; for about 65 per cent control in Sweden, degradation could be up to a factor of 2.42 bringing emissions back up to precontrol levels. For the EEC case, it was decided to use a degradation factor of 1.42 which represents an average value for early controls (up to about 50 per cent) in the U.S.

In order to better represent best and worst case possibilities, three cases are differentiated: (1) strict compliance of new models with standards; (2) an average degradation between cases (1) and (3), taking into account good maintenance of park of the car population (emission factors represented in Table 4.7); (3) use of the 1.42 degradation factor (see Table 4.10).

Within the EEC, the latest standard introduced is the amendment of October 1979 requiring an additional 12.5 per cent control for HC emissions. No decision has yet been taken for future years; it has, however, been assumed here, along the lines of a German proposal, that a new amendment requiring an additional 40-45 per cent control will be adopted for model year 1984. The development of the technology by 1984 should allow this level of control without requiring the use of catalytic devices and unleaded fuel.

Under such conditions, the average HC emission factor would be 28.5 kg/Toe in 1985, and 22.8 kg/Toe in 1990, assuming similar degradation.

For non-highway vehicles, the value developed from the U.S. data (24 kg/Toe) has also been used here.

The EEC emission factors have been used for all countries that do not have control programmes of the U.S. or Japanese type.

Conclusions

The introduction of catalyst emission control technology in motor vehicles enabled several countries to set up stringent emission regulations and to obtain significant emission reductions. Development of the non-catalyst control technology is also likely to produce
_____
* Chapter 4 of Reference (10).

marked reductions. It is, however, still too early to determine whether comparable reductions will be attainable and whether overall costs will be lower for a given low-emission level.

In this respect, it is worth noting that in a recent report from the Australian Committee on Motor Vehicle Emissions (25) it is stated that for Australian conditions a more cost- and energy-effective approach would be to try to achieve a 0.93 g/km HC standard using unleaded technology, rather than a 1.75 g/km HC standard using the leaded technology.

Appendix
to Chapter IV

DETAILED CALCULATIONS OF US AND EEC HIGHWAY VEHICLE EMISSION FACTORS IN KG/TOE

Tables presented below have been included in this appendix to explain more clearly how the US and EEC highway vehicle emission factors in kg/Toe were developed and to show the evolution of these factors with time and under different sets of assumptions.

Table 4.8 presents HC emission factors for the US and EEC respectively, for three cases: (1) strict compliance of new models with standards; (2) taking into account partly degradation of control equipment, partly good maintenance; (3) reflecting average in-use emission rates (5th year of car operation, 80,000 km) as presented in Table 4.3.

In Table 4.8, case 2 was calculated based on US emission estimates (EPA-4501 2-78-052) (23), Table 2.4 in Chapter II of this document, OECD Energy Balances (24), and Secretariat forecasts; base year for calculations was 1970, since no pre-1970 estimates were available in the references. To calculate precontrol levels in kg/Toe, the first step was to determine what fraction of different model year cars, and thus controls, was included in the 1970 value of Table 4.8. This was done using the percent reduction from control values in g/km given in Table 4.9 and assuming a 10 per cent yearly replacement of the car population. It resulted that the 1970 value of Table 4.8 represented 89 per cent and 91 per cent of the precontrol values for case (1) and (3), respectively. This allowed to determine column 3 for both cases in Table 4.9, which presents emission factors for each model year; these were then used to calculate columns 1 and 3 in Table 4.8, assuming 10 per cent of the car population being replaced each year, based on the following equation:

$x = (x - 1) \times 0.9 + $ (model year E.F.) $\times 0.1$
where x = average highway EF for a given year
$x - 1$ = EF of previous year

In Table 4.10, case (1) was calculated based on the same assumption of strict compliance for new models with the standard; case (3) assumes a degradation factor of 1.42 (from US pre-catalyst cars).

All other assumptions are the same as for Table 4.8. Case (2), however, represents a 50/50 weighted average between cases (1) and (3), i.e. a 1.21 degradation factor, since no emission data were available for Europe for each respective year. Table 4.11 presents model year emission factors for Europe based on the same assumptions as Table 4.9 for the US. Base year for calculations was 1970 (i.e. precontrol).

It is noteworthy that precontrol values for the US and EEC based on kg per Toe are almost equal (41.1 and 41.2). Case (2) of above calculations was used as the basis for the emission estimates in the Annex.

Table 4.8

UNITED STATES HIGHWAY VEHICLE HC EMISSION FACTORS

Kg/Toe

Year	Case 1	Case 2	Case 3	Standard (g/km)	Deviation Factor of Case 3 over Standard
Precontrol	41.7	41.1(a)	40.6	(5.4)	(0.92)
1968	40.4	39.8(a)	39.3	3.7	0.92
1969	39.2	38.7(a)	38.2		
1970	37.1	37.1	37.1	2.4	1.42
1971	35.2	35.5	36.1		
1972	33.2	34.3	35.3		
1973	31.3	31.9	34.5	1.9	1.79
1974	29.7	30.3	33.8		
1975	27.4	28.2	31.7		
1976	25.4	27.9	29.9		
1977	23.5	26.8	28.2	0.93	1.72
1978	21.9		25.7		
1979	20.4		23.4		
1980	18.6		21.4		
1981	16.9		19.5		
1982	15.4		18.0		
1983	14.1		16.6		
1984	12.9		15.2		
1985	11.8	13.7	14.1	0.25	(1.72)
1986	10.8		13.0		
1987	9.9		12.1		
1988	9.1		11.2		
1989	8.4		10.4		
1990	7.8	10.0	9.7		

a) Pre-1970 values for case 2 are the arithmetic average of cases 1 and 3.

Values in parentheses are estimates.

Table 4.9

UNITED STATES MODEL YEAR EMISSION FACTORS (EF)

kg/Toe

Year	Case 1			Case 3		
	g/km	%	Model year EF	g/km	%	Model year EF
Precontrol	5.4	100.0	41.66	5.0	100	40.62
1968	3.7	68.5	28.54	3.4	68	27.62
1970	2.4	44.4	18.50	3.4	68	27.62
1972	1.9	35.2	14.66	3.4	68	27.62
1975	0.93	16.7	6.96	1.6	32	13.00
1980	0.25	4.6	1.92	0.43	8.6	3.49

Table 4.10

EEC HIGHWAY VEHICLE HC EMISSION FACTORS

kg/Toe

Year	Case 1	Case 2	Case 3	Standard (g/test)	Deviation Factor of Case 3 over Standard
Precontrol	41.2	41.2	41.2		
1971	41.2	41.2	41.2		
1972	39.9	40.5	41.0		
1973	38.7	39.8	40.9	9.4	1.42
1974	37.6	39.2	40.7		
1975	36.6	38.6	40.6		
1976	35.3	37.6	39.9		
1977	34.2	36.8	39.3	8.0	1.42
1978	33.1		38.8		
1979	32.2		38.3		
1980	31.0		37.4		
1981	30.0		36.7	7.1	1.42
1982	29.1		36.0		
1983	28.3		35.4		
1984	26.6		33.6		
1985	25.1	28.5	31.9		
1986	23.8		30.4		
1987	22.6		29.1	4.0	1.42
1988	21.5		27.9		
1989	20.6		26.8		
1990	19.7	22.8	25.8		

Table 4.11

EEC MODEL YEAR EMISSION FACTORS (EF)

kg/Toe

Year	Case 1			Case 3		
	g/test	%	Model year EF	g/test	%	Model year EF
Precontrol	(13.9)(a)	100	41.20	(13.9)(a)	100	41.2
1972	9.4	67.70	27.89	13.3	95.7	39.4
1976	8.0	57.55	23.71	11.4	82.2	33.9
1980	7.1	50.36	20.75	10.1	72.7	29.9
1984	(4.0)(a)	28.71	11.83	(5.7)(a)	41.0	16.9

a) Figures in parentheses are estimates.

Chapter V

OIL REFINERIES: EMISSIONS AND CONTROLS

Major hydrocarbon emission sources from refineries were described in Chapter III. These sources were grouped according to four stages in refinery operations: processing, effluent water treatment, storage, and loading and unloading. The control technologies available for the first two emission source categories are examined here while the latter two are discussed in the gasoline distribution and marketing section.

When costs are given for control technology, they are capital costs. It is, however, important to note that savings from recovered products will often exceed annual costs. In other words, introduction of control technology will be beneficial not only for environment, health, and safety reasons, but also on economic grounds.

A. PROCESSING AND ASSOCIATED EMISSIONS

Leakage is the most significant cause of emissions during processing operations which implies that good maintenance and housekeeping practices are the primary and one of the most cost-effective ways to reduce HC emissions. These practices, however, may not be sufficient to reduce emissions to an acceptable level in all cases, and control techniques are available for major sources.

<u>Pressure relief valves</u>: Leakage can be prevented by good maintenance or by the installation of rupture discs under the relief valve. These discs (made of steel or carbon) will rupture at a pressure slightly below the set pressure of the relief valve and they have the added advantage of protecting the valve seat from corrosive compounds. Vapour recovery systems can be used to collect vapours blown down from relief valve systems. The average total quantity of volatile organics leaked from refinery relief valves has been estimated by the U.S. EPA (4) at 1.1 kg per day per valve and 1975 costs for rupture discs are given as $1,000 to $1,500 per installation.

<u>Pumps and compressors</u>: According to EPA estimates (1), on an overall refinery basis, hydrocarbon emissions from pumps amount to 50 g/m^3 refinery feed, and from compressors to 14 g/m^3 refinery feed. Emissions from centrifugal pumps can be reduced by 33 per cent by replacing packed seals with mechanical seals. Good maintenance of seals and the use of dual seals with inert fluid or venting to a vapour recovery system can also minimise emissions from pumps and compressors.

The cost of installing a mechanical seal on an existing pump, including a cooler, materials, and labour, is about $2,000 to $2,500* but it will be less if a cooler is not required.

<u>Valves and flanges</u>: Leaks from valves and flanges (and other connecting devices) are normally minimised through regular inspection and good maintenance. The degree of maintenance will depend on a number of factors including temperature and the nature of the products handled, and frequency with which valves are operated and connecting devices assembled and disassembled. The average leak rate from valves is 0.08 kg/10^3 litres refinery feed, according to EPA estimates.** No estimate is available for flanges.

<u>Flares</u>: Flares are mostly used as safety devices but, although their operation reduces hydrocarbon emissions from refineries, they may give rise to other pollutants such as SO_x, NO_x, CO and smoke, as well as visual or noise problems in populated areas. It is often difficult to maintain optimal combustion in a flare, and hydrocarbons are emitted as a result.

Vapour recovery systems can provide a safe and economically attractive alternative when combined with a flare system. The recovered vapour can be reused in the refinery fuel system if recovery of the product itself is not practical or economical.

<u>Blowdown systems</u>: Emissions from uncontrolled refinery blowdown systems have been estimated to be as much as 1.66 kg/m^3 feed.*** Vapours vented to blowdown systems come from all processing units in a refinery. To control these vapours, an extensive system separating discharges into different pressure cuts may be required before venting to a vapour recovery unit. Emissions from controlled blowdown systems have been estimated at 20 g/m^3 of refinery capacity.***

<u>Vapour recovery systems</u>: Vapour recovery systems can be used to control emissions from a large number of refinery emission sources. These systems are based mainly on refrigeration and compression/absorption type units. Vapour recovery plants are normally designed to control about 90 per cent of a fully saturated hydrocarbon vapour.

---

* Reference (4) pp. 122-123.
** Reference (4) p. 119.
*** Reference (4) p. 131.

The cost of vapour recovery units depends on capacity and other factors. Information on costs for gasoline recovery is given in the gasoline distribution and marketing section.

<u>Catalyst regenerators</u>: HC emissions from regenerators depend upon charge and type of cracker. For fluidised bed catalyst cracker regenerators, they average 630 g/m^3 fresh cracker feed; and for moving bed catalyst cracker regenerators, 250 g/m^3 fresh cracker feed. By incinerating regenerator flue gas in CO boilers, which is common practice in Europe, HC emissions can be reduced to 0.57 g/m^3 of feed; valuable thermal energy is thus recovered. Regenerators of recent design operate at higher temperatures, emitting very little CO and HC. Incineration in a process heater box or smokeless flares can also be used to reduce HC emissions from regenerators. However, in all but small refineries, the cost of CO boilers is more than compensated by the fuel savings from heat recovery. Capital costs for a catalytic cracker, processing 4,930 m^3/day, are $3.1 million for a new unit and $3.7 million for retrofit.*

B. EFFLUENT WATER TREATMENT

Studies of refineries in Los Angeles County indicate that average HC emissions from sumps, drains, and American Petroleum Institute (API) separators are 2,700 kg/day ** (ranging from 30 g/m^3 to 600 g/m^3 capacity). Another U.S. estimate set average emissions from refinery waste water systems at 0.3 kg/m^3 refinery feed.**

There are two ways of reducing emissions from oil-water separators. The fist is to limit the amount of oil products leaking from refinery processes and sent to the oil-water separator; the second is to enclose the separator. Enclosure can be done with either a floating roof, which will limit vapour space and hence formation of explosive mixtures, or a fixed roof which will require venting to a blow-down or vapour recovery unit.

Hydrocarbon emissions from API separators can be reduced to 3 g/m^3 capacity by means of floating roofs.*** With fixed roofs and blow-down or vapour recovery, emission reduction efficiency would also reach around 90 per cent but costs would be greater than for floating roofs.

Capital costs for floating roofs on API separators range from $27,800 to $167,300 for refinery size from 1,590 m^3/day to 31,800 m^3/day.

---

* Reference (4) pp. 150-154.
** Reference (4) p. 141.
*** Reference (4) p. 142.

C. OVERALL EMISSIONS FROM REFINERIES

A survey of refinery emissions was recently conducted in Europe. The results, reported by CONCAWE (26), indicate that HC emissions from European refineries are lower than expected.

On the basis of the traditional API/EPA emission factors, process equipment emissions had been estimated at 0.026 per cent wt. of crude throughput. However, according to the European survey, some of these factors overestimate losses from modern equipment, and results indicate that process losses are probably 4 to 10 times lower, i.e., 0.003 - 0.007 per cent wt. of crude throughput.

The evaporation loss from separators, according to a widely accepted correlation, is about 7-8 per cent wt. of the incoming oil. For a simple refinery separator, emissions will not be greater than 0.01 per cent wt. of the refinery crude throughput. The calculated loss for a complex refinery would be about 0.024 per cent wt. of crude throughput.

These losses may be reduced by minimising oil flow to the separators and by reducing high effluent temperatures. Installation of a cover to prevent vapour escape would also reduce losses. In this case, losses would be reduced to about 0.002-0.005 per cent wt. of crude throughput.

Results reported by CONCAWE gave an average emission loss from tankage for 50 European refineries of about 0.02 per cent wt. of the crude oil distillation throughput. Two-thirds of this loss occurs from fixed-roof tankage, although such tankage represents less than one-tenth of the total installed capacity on volatile HC duty in Europe.

For future conditions it has been assumed, on the basis of results reported by CONCAWE, that processing emissions in Europe will be reduced to 0.005 per cent wt. of crude throughput as a result of better maintenance. Evaporation loss from separators can be reduced to 0.010 and 0.005 per cent wt. of crude throughput for 1985 and 1990, respectively, with the gradual introduction of floating covers. Finally, tankage emissions are not expected to change considerably in the future, since 90 per cent of tankage capacity on volatile HC duty in Europe is already of the floating-roof type (see Table 5.1). This implies, however, that additional floating-roof capacity might have to be installed or retrofitted, as the demand for volatile products increases.

It should be noted that Finland also reported to the Secretariat an emission factor for refineries of 0.4 kg/Toe under present conditions.

In the U.S., it is estimated that total refinery emissions are approximately 0.25 per cent wt. (ranging from 0.1 to 0.6) of crude

Table 5.1

TYPICAL HYDROCARBON LOSSES FROM EUROPEAN REFINERIES
as % wt. of crude throughput

	1977	1985	1990
Processing	0.010	0.005	0.005
Separators	0.017	0.010	0.005
Storage	0.020	0.020	0.020
Total	0.047	0.035	0.030

throughput (2.5 kg/Toe). This is significantly higher than the European estimate. Canada reported a refinery emission factor of 0.41 lb/bbl crude received (1.17 kg/Toe). It is not known, however, whether this figure covers only process emissions or whether it includes storage and thus it was difficult to use it.

According to U.S. EPA estimates, HC emissions from refineries will decrease by 30 per cent in 1985 and 1990 from 1975 levels. Taking into account the growth in oil consumption, emission factors for 1985 and 1990 are then 1.31 kg/Toe and 1.20 kg/Toe, respectively.

Because the Australian, Canadian, and U.S. refinery operating practices are fairly similar, the U.S. emission factor has also been used for these other two countries while the European factor has been used for all other OECD countries.

Chapter VI

GASOLINE DISTRIBUTION AND MARKETING: EMISSIONS AND CONTROLS

A typical gasoline distribution system consists of storage at the refinery, transportation by sea or land to a storage terminal and then distribution to service stations. In the U.S., there is sometimes an additional step whereby gasoline is transferred from the refinery to a large bulk plant, then to a local depot, and finally to the service stations. However, in all OECD countries, some gasoline goes directly from the refinery to service stations.

Gasoline distribution and marketing involve a number of operations (transfer and storage) which are potential sources of HC emissions. Even though these losses are generally only a small fraction of the total product throughput, they should be minimised or eliminated for economic, environmental, safety, and energy-saving reasons.

Some countries have legislated regulations to limit these emissions, particularly for health and safety reasons. However, because of the rapid increase in energy prices, the oil industry has often introduced control technology, even when there was no standard to be met. Control technologies are available for both storage and transfer operations. These technologies, their impact on emissions, and their costs are reviewed below.

A. TERMINAL STORAGE EMISSION CONTROLS

The most common technique to prevent emissions from depot storage is floating roofs or covers, which reduce evaporative losses by minimising the vapour space in the tank. There are several types of floating roof or cover, namely the pontoon type, the pan type, the double deck type, and the self-buoyant sandwich type.

The floating roof is used mostly on new tanks where weather conditions allow the use of conventional open tanks. According to American Petroleum Institute (API) estimates, the use of floating-roof tanks can reduce emissions by 90-95 per cent when compared to conventional fixed-roof tanks in the U.S.

Internal floating covers are generally of light construction to allow retrofitting to existing fixed-roof tanks. According to estimates by CONCAWE, the efficiency of floating covers is approximately 80 per cent when compared to a fixed-roof tank. It should, however, be noted that the pressure/vacuum (P/V) valve setting on conventional fixed-roof tanks in Europe is generally much higher than in the U.S. (26 mbar or 56 mbar compared to 6 mbar, respectively), which influences considerably the uncontrolled emissions from fixed-roof tanks. Furthermore, it is recommended practice, and in some countries compulsory, for safety reasons, to ensure ventilation of the vapour space in floating roof tanks to prevent the formation of a flammable mixture.

Other factors may have a significant impact on emissions from storage tanks. More specifically, if the seal of the floating roof or cover is in poor condition, the emission control efficiency can be reduced to 15 per cent. Also, emissions can be influenced by the reflectivity of the tank surface coating; hence, careful selection of paint colour and good paint condition are essential.

There are other control technologies for terminal storage tanks, such as vapour balancing, variable vapour space tanks, and tank insulation but these techniques have had limited use to date. It appears that vapour balancing has not been applied in Europe because of the increasing use of internal floating covers and the absence of vapour recovery equipment at refineries. Variable vapour space tanks best suited for seasonal or strategic storage have only been used for gasoline in the U.S., Australia and Canada. Finally, tank insulation would be beneficial only when compared to other techniques with specific climatic conditions as well as very slow tank turnover.

As floating roofs or covers are the most commonly used technology, emission factors and control costs in this report are given for this control measure only.

For marketing terminals, the average emission based on typical European fixed-roof tankage, and assuming an annual true vapour pressure (TVP) of 350 mbar, is 0.09 to 0.23 for winter and summer; for floating-roof tanks, the average emission is 0.03 per cent vol., which represents about 80 per cent control efficiency.

For a 5,000 m^3 fixed-roof tank (20 m diameter, 16.8 m high), the cost of retrofitting an internal floating cover has been estimated at $29,100 by CONCAWE (26); this includes cleaning, degassing, and preparation work for the tank. In the U.S., the cost for a tank of the same capacity would be approximately $20,000, without cleaning costs, according to EPA estimates (4). Costs for cleaning and degassing for a fixed-roof tank were reported as ranging from $5,500 to $8,000; this gives an overall cost similar to the European estimate.

B. DESPATCH EMISSION CONTROLS

Despatching of gasoline usually takes place twice in a distribution system. Firstly, gasoline is transported from the refinery to the storage terminal by waterway (barges or tankers), or by land (rail or pipeline); transportation emissions from pipelines are nil or negligible under normal conditions and consequently no control measure will be considered here. Secondly, gasoline is transported by tank trucks from the terminal to service stations.

1. Barges and Tankers

Emissions from barges and tankers arise essentially during filling operations. These emissions are due to the preloading vapour concentration in the transport compartment, and to the type and mode of operation of the filling device. Maximum emissions are obtained when transport compartments which had not been gas-freed are ventilated to the atmosphere and subsequently filled by splash loading. With tankers, emissions also originate from ballasting operations unless tankers are equipped with segregated ballasts.

The best method of reducing emissions during filling is to use bottom loading. This prevents splash and turbulence and hence reduces emissions by about 80 per cent from splash loading. Modern design vessels normally have bottom loading and segregated ballasts.

However, with bottom loading, marine loading losses are still 0.018 per cent vol. for ships, and 0.010-0.018 per cent vol. for barges if vessel tanks are gas-free at the start of loading - these losses rise to 0.34 per cent vol. for tankers and 0.057 per cent vol. for barges if the tanks have not been gas-freed, according to CONCAWE estimates (26). The corresponding figures from EPA estimates (4) are: 0.020 per cent vol. for gas-free tankers, 0.023 per cent vol. for gas-free barges, and 0.048 per cent vol. and 0.079 per cent vol., respectively, for tankers and barges that have not been gas-freed. It should be remembered when comparing these figures that TVP and loading rate of gasoline in the U.S. are generally higher than in Europe, which may account for the differences.

To reduce these emissions vapours can be collected and sent to either a vapour recovery or combustion unit.

A vapour recovery unit of the compression/absorption type normally has a control efficiency of about 80 per cent while that of a refrigeration type unit is about 90 per cent. From CONCAWE estimates (26), for a typical case the capital cost of a vapour recovery plant would be \$216,500 for 360 m^3/h capacity, and \$345,500 for 660 m^3/h capacity, plus a cost of \$54,100 for the vapour return pipework. The EPA (4) has estimated that the average capital cost in the U.S. would be \$1,000,000 for 1600 m^3/h capacity; this is comparable to the CONCAWE estimates when reduced to the same scale. Additional costs may be incurred because of vessel modifications; these have been estimated at $0.35 million per tanker and $67,000 per barge (EPA), and $45,900 to $119,000 per barge, depending on type and size (CONCAWE).

Annual average operating costs have been estimated in the U.S. (27) at $15/$10^3$ barrels throughput.

Hydrocarbon vapour/air mixtures from transportation compartments during loading can also be burned in a combustion chamber. However, the use of these units is limited because of a number of factors including irregular vapour supply and concentration, the need to add fuel to sustain combustion, and safety problems. Capital costs were reported by CONCAWE as $125,000 for a 335 m^3/h of vapour unit (Europe), and $105,000 for a 170 m^3/h vapour unit (U.S.); in both cases, the cost of the return pipework and loading arms must be added.

2. Rail Cars and Tank Trucks

As in the case of barges and tankers, emissions from rail cars and tank trucks originate essentially from loading operations. The mechanisms of vapour generation and the factors affecting loss rate are the same as those described for marine loading.

Splash loading, top submerged loading, and bottom loading systems are used. Submerged and bottom loading methods reduce emissions by 40 to 60 per cent from splash loading. Emissions from submerged loading and splash loading are reported by CONCAWE (26) as: 0.061 per cent vol. and 0.13 per cent vol., respectively (from rail cars); and 0.055 per cent vol. and 0.15 per cent vol., respectively (from tank trucks). In the U.S. (4) these emissions from tank trucks and rail cars are reported as: 0.6 kg/10^3 litres and 1.4 kg/10^3 litres in AP42. After conversion into per cent volume, U.S. emissions factors are somewhat higher than the CONCAWE estimates; this may be due to the higher TVP, loading rate and preloading concentrations assumed in the U.S. cases.

Emissions by top-submerged loading can be limited, to a certain extent (10-20 per cent), by the use of a slow opening valve, or the replacement of top-submerged loading by bottom loading. The cost for a bottom-loading installation, including rail car or truck modification, would range from $13,000 to $20,000 according to CONCAWE estimates (26).

A much more cost-effective technique to control loading emissions from rail cars and tank trucks is vapour balancing. This method is used in the U.S. but rarely, if at all, in Europe. With this technique, vapours emitted and displaced during loading are collected in a vapour return pipe and sent to either a fixed-roof tank, or to a vapour-recovery unit if only floating-roof tanks are available. The efficiency of vapour balancing is 90 to 98 per cent, but the actual quantity of vapour recovered will depend on whether vapour balancing is also practised between the point of delivery and the tank truck/rail car.

According to CONCAWE (26), the cost for vapour balancing would range from \$8,300 to \$17,500 for a truck or rail car loading rack, depending on whether retrofitting is practised; for vapour recovery plant, the cost would range from \$216,500 (360 m^3/h) to \$345,500 (660 m^3/h).

U.S. EPA estimates (4) for vapour recovery systems range from \$47,000 (capacity of 75.7 m^3/day) to \$185,000 (capacity of 950 m^3/day). It is not known whether the U.S. estimates include loading rack modification costs.

C. SERVICE STATION EMISSION CONTROLS

Most HC emissions at service stations originate from two operations - underground storage tank filling and vehicle refuelling. Breathing emissions from service station tanks are normally very small, since tanks are buried and therefore not subject to much temperature variation.

Estimates of emissions by EPA (4) for submerged filling and splash filling of service station tanks are 0.14 per cent vol. and 0.23 per cent vol., respectively. CONCAWE estimates (26) are 0.16 per cent vol. for European conditions; it is assumed that this figure represents submerged filling operations.

Displacement emissions from service station tank filling can be controlled by vapour balancing with 90-100 per cent efficiency. Vapours from the tank are then returned to the delivery truck during tank filling. Estimates of controlled emissions are 0.007 per cent vol. (EPA) and 0.016 per cent vol. (CONCAWE).

Capital costs for retrofitting service stations with vapour balance have been estimated at \$6,000 for a 120 m^3/month U.S. station (EPA), and \$4,200 for a 40 m^3/month European station (CONCAWE). Tank modification for vapour return would cost an additional \$5,000 to \$9,000 per truck (26).

Emissions during automobile refuelling are caused by vapours being displaced from the vehicle fuel tank. These emissions have been estimated at 0.178 per cent vol. (EPA) and 0.175 per cent vol. (CONCAWE). Vapour balancing can be used to control these emissions with an efficiency of 70 to 80 per cent. This efficiency may reach 90 per cent with a vacuum-assist system and excess vapour control. Vapour balancing will also prevent the evolution of vapour in the service station tank (0.16 per cent vol.), and the withdrawal emission (0.01 per cent vol.) since no air will be drawn into the tank. However, vapour balancing for service station applications is not yet used in Europe, and is currently under study in the U.S., because operational and maintenance problems still need to be solved.

Capital costs for a typical European service station (40 m^3/month), comprising two tanks and four dispensers, have been estimated by CONCAWE (26) at $3,900 without a vacuum-assist system, and at $11,400 with this system. Additional costs may arise from the need to incorporate safety equipment and to modify the car tank inlet.

D. THE EUROPEAN SITUATION

Table 6.1 is a summary of HC emissions from a typical European distribution system. Four scenarios are considered:

1. Fixed-roof tankage in gasoline terminals and no vapour balancing;
2. Fixed-roof tankage in gasoline terminals and vapour balancing throughout the whole distribution system (i.e., from the automobile fuel tank back to the refinery);
3. Internal floating covers in gasoline terminals, and no vapour balancing;
4. Internal floating covers in gasoline terminals, and vapour balancing from the automobile fuel tank back to terminal despatch.

On the basis of these four scenarios, gasoline distribution and marketing emission factors for European countries have been developed for 1977, 1985, and 1990, with the following assumptions:

a) for 1977: a 20 per cent wt. gasoline yield on refinery crude oil throughput; 50/50 fixed/floating roof tankage in marketing terminals; 50/50 land/marine distribution.
b) for 1985: a 25 per cent wt. gasoline yield on refinery crude oil throughput; 20/80 fixed/floating roof tankage in marketing terminals; 50/50 land/marine distribution.
c) for 1990: a 25-30 per cent wt. gasoline yield on refinery crude oil throughput; 10/19 fixed/floating roof tankage in marketing terminals; 50/50 land/marine distribution.

As can be seen on Table 6.2 the assumption that the barrel will become lighter over the next decade in Europe has a very significant impact on potential HC emissions from the gasoline distribution system. Even without taking into account an increase in refinery crude throughput, emissions will increase by a factor of 1.8 by 1990, if

Table 6.1

HYDROCARBON EMISSIONS IN A EUROPEAN DISTRIBUTION SYSTEM

% loaded liquid volume

Stage	Fixed-roof tank +20/-5 mbar		Internal-floating cover	
	Vapour balancing		Vapour balancing	
	None	Stage g → stage a + recov. at stage a	None	Stage g → stage d + recov. at stage d
a. Refinery Despatch	0.061	0.016	0.061	0.061
Terminal storage:				
b. displacement	0.14	0.014	0.028	0.028
c. breathing and withdrawal	0.02	0.02	0.003	0.003
Terminal Despatch:				
d. vehicle loading	0.055	0.016	0.055	0.016
Service Station:				
e. displacement	0.16	0.016	0.16	0.016
f. breathing and withdrawal	0.01	0.01	0.01	0.01
Motor Vehicle Refuelling:				
g. displacement	0.175	0.044	0.175	0.044
Total	0.621	0.136	0.492	0.178

Efficiency of vapour return: 90% of displacement for depot and station receipt; 75% for vehicle refuelling.
Efficiency of internal floating cover: 80% of displacement; 85% of breathing and withdrawal emissions.
Efficiency of recovery: 90%.

Source: Reference (26).

the only control measure taken is the slow introduction of internal floating covers on terminal tankage. However, if vapour balancing is introduced as well, emissions by 1990 will decrease by 30 per cent from 1977 levels, assuming 1977 refinery crude throughput.

Since no legislation, existing or planned, requires the use of vapour balancing in Europe, the values corresponding to the "no vapour balancing" case have been used for the emission forecasts. These values have also been used for other OECD countries, except Australia, Canada, and the U.S.

Table 6.2

EVOLUTION OF DISTRIBUTION
EMISSION FACTORS IN EUROPE

in % wt. of refinery crude throughput

kg/Toe

	Vapour balancing	
	with	without
1977		0.09 (0.9)
1985	0.05 (0.5)	0.15 (1.5)
1990	0.06 (0.6)	0.16 (1.6)

E. THE U.S. SITUATION

According to U.S. EPA estimates, overall emissions from gasoline distribution and marketing amount to 0.30 per cent wt. of refinery crude throughput under present conditions. A detailed breakdown of this estimate on the same basis as the European one is, however, difficult because some data are lacking. On the other hand, for comparison purposes, if CONCAWE emission factors (0.621 and 0.492 from Table 6.1) are used in combination with U.S. conditions (50 per cent gasoline yield and 40/60 fixed/floating roof tankage), the emission factor obtained is 0.22 per cent wt. of refinery crude throughput. The 0.08 difference can be explained by a number of factors such as gasoline TVP, P/V valve settings and loading rates, as well as the fact that according to recent tests, the equations used in the U.S. emissions tend to overestimate losses.

For 1985 and 1990, the U.S. EPA estimates that emissions in this sector will increase by 35 per cent and 55 per cent, respectively, over 1975 levels. Taking into account the increase in oil consumption, emission factors for 1985 and 1990 become respectively, 0.30 per cent wt. and 0.32 per cent wt. of refinery crude throughput. Although these factors are very similar to the 1970s situation this does not necessarily mean that no control will be introduced in the future. Even though refineries in the U.S. already have a high gasoline yield on crude, a further increase of this yield may conceal in the emission factors the benefit that would be obtained from introducing controls.

In this exercise, the U.S. values have been considered to be applicable to Canada and Australia, since the refining pattern and gasoline distribution conditions of these two countries are similar to those of the U.S.

Chapter VII

EMISSION LEVELS FROM OTHER HYDROCARBON SOURCES

The previous three chapters have discussed HC emissions and controls from the main source - the gasoline cycle. This source category, including refinery emissions, gasoline distribution and marketing, and motor vehicle emissions, represents approximately 50 per cent of overall HC emissions in most OECD countries. The remaining 50 per cent are emitted by a large number of generally small sources which differ in nature as well as type.

The implementation of emission controls on these sources would often be difficult, and costs could be very high, even though technologies are available. Consequently, a detailed review of control technologies and costs for these sources has not been included with the following description of emission factors.

The following sources were used to derive the emission factors discussed in this chapter:

For industry in general:

United States: AP42 Compilation of Air Pollutant Emission Factors - Third Edition, August 1977.

Italy and Canada: National response to an OECD questionnaire.

United Kingdom: Emissions Inventory for Hydrocarbons in the United Kingdom, K.A. Brice and R.G. Derwent, Atmospheric Environment 1978.

Germany: Materialien zum Immissionsschutzbericht 1977, Umweltbundesamt, Berlin.

For the petroleum industry:

- AP42 (see above) and EPA-450/2.78.022 (Control techniques for VOC emissions from stationary sources) for the United States.
- Report no. 4/78 (Hydrocarbon emissions from gasoline storage and distribution systems, CONCAWE, The Hague, 1978) for Europe.

1. ELECTRICITY GENERATION

Coal

	Hardcoal			Lignite		
	kg/TJ	kg/MT	kg/Toe	kg/TJ	kg/MT	kg/Toe
FRG	3.4	0.14	0.14	3.4		0.14
Italy(a)		0.3 - 0.5	0.82-0.68		0.15	0.60
United Kingdom(b)		0.15	0.28			
United States		0.15	0.23		0.5	0.82
Canada(c)		0.12-0.17	0.26-0.25			

a) Range for national and foreign coal respectively.
b) Reference (3).
c) Range for sub-bituminous and bituminous.

On the basis of the above table, it was decided to use 0.15kg/MT for hardcoal which gives an average of 0.21kg/Toe for all OECD countries. The same factor on a Toe basis has been used for lignite because its combustion characteristics are similar to bituminous coal, even though national emission factors tend to be higher.

Oil

	kg/TJ	kg/MT	kg/10^3 litre	kg/Toe
FRG	6.8			0.28
Italy		0.25		0.26
United Kingdom(a)			0.25	0.27
United States(b)			0.12	0.14
Canada			0.25	0.27

a) Reference (28).
b) Reference (3).

As can be seen from the above table, the U.S. EPA recently reduced their stated VOC emission factor for oil combustion at stationary sources on the basis of better and more recent information. It was decided to use this new factor here (i.e. 0.14 kg/Toe).

Gas

	kg/$10^6 m^3$	kg/TJ	kg/Toe
FRG		0.45	0.02
United States	16		0.02
Canada	16		0.02

For gas combustion in power stations, 0.02 kg/Toe has been used.

2. FUEL PRODUCTION

VOC emissions in this sector include emissions from processes associated with the extraction and/or production of primary energy, i.e. coal mining, oil and gas production.

Coal

The following factor is an average of the industrial coal combustion emission factors as given in (3):

$$0.325 \text{ kg MT}^{-1} \quad (0.46 \text{ kg Toe}^{-1})$$

Oil

VOC emissions arise from oil production as well as from crude storage at the production site. From available information, it appears that current U.S. practice is to use fixed roof tanks for this storage. Two sets of emission factors were made available to the Secretariat:

	U.S. EPA	CONCAWE
Crude Production	0.48	
Crude storage	1.29	
Total	1.77 kg/Toe	1.00 kg/Toe

The CONCAWE factor has been used for all countries except the U.S., where the EPA factor has been applied.

Gas

Natural gas production processes can be considered as a large VOC emitter, however, about 90 per cent of the emissions are methane. The emission factor used for natural gas production processes is a U.S. EPA estimate:

$$190 \text{lb}/10^6 \text{ standard cubic feet } (3.6 \text{ kg/TOE})$$

The production of natural gas is often associated with the production of liquified natural gas (LNG). However, according to production statistics, LNG production is only significant in the U.S. and Canada where they represent 10-13 per cent of natural gas production. The following emission factor from U.S. EPA has been used for these two countries:

$$13.4 \text{lb}/10^3 \text{ gallons } (2.46 \text{ kg/Toe})$$

3. INDUSTRY SECTOR

The diversity and complexity of this sector, both within each individual country and within the OECD as a whole, has lead to a number of assumptions in calculation of VOC emissions. Three classes of emission source were defined: combustion, chemical and petrochemical processes, and solvent evaporation.

Combustion

For coal combustion, it was decided to use an average of the industrial combustion emission factors given in (3):

$$0.325 \text{ kg/MT} \quad (0.46 \text{ kg/Toe})$$

For oil and gas, emission factors from AP42 have also been used:

oil - $0.12 \text{ kg}/10^3$ litre $\quad (0.14 \text{ kg/Toe})$
gas - $48 \text{ kg}/10^6 \text{ m}^3 \quad (0.06 \text{ kg/Toe})$

Chemical and petrochemical processes

On the basis of consumption data for naphtha and liquified gases and VOC emission inventories for these industries, the following average emission factor was developed:

$$40 \text{ kg/Toe}$$

Solvent evaporation

This sector has been included under a different approach and is examined below in a separate section.

4. RESIDENTIAL/COMMERCIAL SECTOR

Coal

	Hardcoal		
	Kg/MT	kg/TJ	kg/Toe
FRG(a)		350-200	14.42-8.24
Italy(b)	0.3-0.5		0.82-0.68
United Kingdom	1.25		1.78
United States	5.78		8.26

a) Range for 1975 and 1980 with control, respectively.
b) Range for national and foreign coal, respectively.

For the U.S., in 1973, bituminous coal emissions represented 87 per cent of all VOC emissions from coal combustion in this sector(29). Using these 1973 coal emission data and coal consumption figures (30), the above emission factor was developed. This factor is practically equal to the average of domestic furnaces and hand-fired units burning bituminous coal: 5.75 kg/MT, according to U.S. literature (3). This factor, on a Toe basis, seems to be in general agreement with the German data. As the relative share of anthracite and bituminous coals is not known for all OECD countries, it has been decided to use an average factor of 8.00 kg/Toe.

	Lignite		
	kg/MT	kg/TJ	kg/Toe
FRG		300	12.4
Italy	0.15		0.60
United States	0.5		0.82

As can be seen in the above table, the range of national factors in this case seems extremely wide (> 20). It should, however, be noted that the United States factor does not specifically apply to the residential commercial area. Furthermore, the U.S. EPA mentions in (3) the similarity between lignite and bituminous coal combustion, which is reflected in the German factor. As no details could be obtained on the Italian factor, it was decided to use here the same approach as for power plant solid fuel combustion and to choose 8.00 kg/Toe.

Oil

	kg/MT	kg/10^3 litre	kg/TJ	kg/m^3	kg/Toe
FRG			15		0.63
Italy	0.35				0.34
Norway(a)				0.35	0.40
United Kingdom(a)		0.35			0.40
United States(b)		0.12			0.14

a) Reference (28).
b) Reference (3).

As already stated, the U.S. EPA revised their oil combustion factor in 1977 after Norway and the United Kingdom conducted their emission inventory. It was decided to use this more recent factor here (i.e., 0.14 kg/Toe).

Gas

	kg/10^3 m^3	kg/10^6 m^3	kg/TJ	kg/Toe
FRG			12	0.55
Italy	0.13			0.15
United Kingdom		128		0.15
United States		128		0.15

It appears that Italy and the United Kingdom have been using the factor given in (3). The German factor, however, is based on measurements carried out in Germany and is somewhat higher than the United States factor. It was decided to use 0.15 kg/Toe.

5. SOLVENT EVAPORATION

This source category includes VOC emissions from all uses of organic solvents, both industrial and non-industrial. Because of the enormous range of uses of these products and the difficulty of obtaining data on present and future production and use, it was necessary to use a different basis for emission estimates and projections.

According to the U.S. literature (5), there is a relationship between solvent use and population. This relationship was developed for countries for which emission data were available. For other countries, the emission factor was extrapolated taking into account GDP per capita. Base year emissions were then projected into the future using GDP growth rates. Results are presented in Table 7.1.

Table 7.1
SOLVENT EVAPORATION
(kg/year/cap)

	1975	1985	1990	2000
United States	29.5	40.4	44.5	55.2
Canada	10.0	13.5	15.6	19.2
Japan	9.5	15.6	19.2	27.0
Australia	14	18.5	21.1	27.4
New Zealand	8.2	10.3	11.2	13.1
Austria	9.5	13.4	15.8	21.5
Belgium	12.2	17.4	21.0	30.2
Denmark	13.4	20.1	24.6	32.3
Finland	10.8	15.7	18.2	24.5
France	12.2	17.5	21.0	29.5
Germany	13.2	18.8	21.4	27.9
Greece	4.4	8.0	10.2	16.7
Iceland	10.9	12.4	15.6	25.3
Ireland	4.7	6.3	7.4	10.0
Italy	5.9	8.6	9.9	13.5
Luxembourg	11.8	17.3	20.0	25.5
Netherlands	11.4	15.3	17.0	21.2
Norway	13.5	20.9	25.0	35.9
Portugal	3.0	4.8	5.9	9.2
Spain	5.5	6.8	7.8	10.8
Sweden	12.1	16.1	17.9	22.4
Switzerland	16.2	19.4	21.8	27.6
Turkey	1.7	2.9	3.5	5.6
United Kingdom	5.5	7.4	8.5	10.9

REFERENCES

1. OECD: Photochemical Oxidant Air Pollution, OECD, Paris, 1975.
2. OECD: Photochemical Oxidants and their Precursors in the Atmosphere - Effects, Formation, Transport and Abatement, OECD, Paris, 1979.
3. U.S. Environmental Protection Agency: Compilation of Air Pollution Emission Factors, 3rd Edition, AP42, August 1977.
4. U.S. EPA: Control Techniques for Volatile Organic Emissions from Stationary Sources. EPA-450/2-78-022, May 1978.
5. U.S. EPA: Air Pollution Engineering Manual, USEPA, AP40, May 1973.
6. Brice, K.A. and Derwent, R.G.: Emission Inventory for Hydrocarbons in the United Kingdom - Atmospheric Environment, Pergamon Press, Oxford, 1978.
7. Black, F: The Impact of Emission Control Technology on Passenger Car Hydrocarbon Emission Rates and Patterns. International Conference on Photochemical Oxidant Pollution and its Control, U.S. Environmental Protection Agency. EPA-600/3-77-001, January 1977.
8. OECD: Automotive Air Pollution and Noise: Implications for Public Policy. (Document) OECD, Paris, 1972.
9. Japanese Environment Agency: Air Pollution and Motor Vehicle Emission Control in Japan, March 1977.
10. OECD: The Cost and Effectiveness of Automotive Exhaust Emission Control Regulations, OECD, Paris, 1979.
11. National Academy of Sciences: Report by the Committee on Motor Vehicle Emissions, Washington, 1974.
12. U.S. EPA: Cost and Economic Impact Assessment for Alternative Levels of the National Ambient Air Quality Standards for Ozone, EPA-450/5-79-002, February 1979.
13. U.S. EPA: Analysis of Technical Views relating to California's Request for waiver of Federal Pre-emption with respect to Exhaust Emission Standards and Test Procedures of 1981 and subsequent model years light-duty vehicles. EPA Emission Control Technology Division, March 1978, p. 21.

14. Umweltbundesamt (UBA): Recommendations to Reduce the Emissions of Pollutants - Motor Vehicle Exhaust Emissions, Federal Environment Agency, Berlin 1976.
15. Japanese Environment Agency: Quality of the Environment in Japan, 1977.
16. U.S. EPA: Comparison of Japanese and U.S. Automotive Emission Standards, EPA, Office of Air and Waste Management, Washington, 1977.
17. U.S. EPA: An Analysis of Alternative Motor Vehicle Emission Standards, EPA, Washington, 1976.
18. U.S. EPA: Tradeoffs Associated with Possible Auto Emission Standards, Washington, 1975.
19. International Permanent Bureau of Motor Manufacturers (BPICA): Increases in Cost and Fuel Consumption of Passenger Cars resulting in CO and HC Emission Limits as from 1980, Geneva, 1976.
20. U.S. EPA: Control Strategy Preparation Manual for Photochemical Oxidant. Office of Air Quality Planning and Standards, EPA Guidelines Series No. 1, 2 - 047, January 1977.
21. Interplant Corp.: Transportation System Management: State-of-the-Art Report for Urban Mass Transportation Administration, U.S. Department of Transport, February 1977.
22. OECD-ECMT: Urban Transport and the Environment, Volume 3, Seminar, Paris, 1979.
23. U.S. EPA: National Air Quality Monitoring and Emission Trends Report. EPA-450/2-78-052, December 1978.
24. OECD: Energy Balances of OECD Countries, OECD, Paris, 1979.
25. Commonwealth Department of Transport: Proposal for the Development of a Long-term National Motor Vehicle Emissions Strategy from the Australian Committee on Motor Vehicle Emissions, March 1980.
26. CONCAWE: Hydrocarbon Emissions from Gasoline Storage and Distribution Systems. Report No. 4/78, December 1978.
27. Busklin, E.C. et al: Background and Information on Hydrocarbon Emissions from Marine Terminal Operations. EPA Project No. 68-02-1319, Task 56, Radian Corp. Texas, November 1976.
28. U.S. EPA: Compilation of Air Pollution Emission Factors. 1st Edition, AP42, 1973.
29. U.S. EPA: Nation Emissions Report, EPA-450/2-76-007, May 1976.
30. OECD: Energy Statistics 1973-1975, OECD, Paris, 1976.

Annex

EMISSION ESTIMATES AND FORECASTS

The emission estimates for 1977 and forecasts for 1985 and 1990 presented here were developed on the basis of the emission factors discussed in the report. It should be noted that in the forecasts for 1985 and 1990 it has been assumed that no further control measures against VOC emissions are taken after 1977, except for the gasoline cycle (including petroleum refining operations, gasoline distribution, and the transport sector).

A few general remarks can be made on these emission estimates. For most European countries, the total emissions are estimated to increase in the future and in some cases practically double over this 13-year period. In the case of Norway, the very significant expansion of its oil and gas production has a dramatic effect on total emissions. It should be remembered, however, that most of these operations will be offshore, and in the case of emissions from natural gas production processes, 90 per cent of these emissions are methane, which implies no impact on photochemical oxidant formation. The estimates for U.K. and Swedish total VOC emissions stabilize after 1985.

In other OECD countries, different trends can be observed. In Australia and the United States, total emissions stabilize after 1977, while in Canada they decrease over the same period. In Japan, however, emissions from industry, solvent evaporation, and gasoline distribution appear to increase dramatically.

AUSTRALIA

1977

VOC Emissions   1000 Metric Tonnes

	Coal	Oil	Gas	Total
Production	1.54	35.68	26.23	63.45
Pet. ref.	*	72.08	*	72.08
Gasoline dist.	*	86.49	*	86.49
Electricity	3.65	0.07	0.02	3.74
Industry	2.01	0.81	0.18	3.01
Transport	*	386.48	*	386.48
Res./com.	7.14	0.43	0.17	7.74
Solvent evap.	*	*	*	203.00
Total	14.34	582.04	26.60	825.99

1985

VOC Emissions   1000 Metric Tonnes

	Coal	Oil	Gas	Total
Production	2.10	16.73	74.22	93.05
Pet. ref.	*	45.06	*	45.06
Gasoline dist.	*	103.20	*	103.20
Electricity	5.52	0.11	0.04	5.67
Industry	0.69	0.95	0.43	2.07
Transport	*	242.40	*	242.40
Res./com.	0	0.77	0.39	1.16
Solvent evap.	*	*	*	303.00
Total	8.31	409.22	75.08	795.61

1990

VOC Emissions   1000 Metric Tonnes

	Coal	Oil	Gas	Total
Production	2.91	15.13	96.30	114.34
Pet. ref.	*	45.60	*	45.60
Gasoline dist.	*	121.60	*	121.60
Electricity	6.47	0.10	0.04	6.61
Industry	1.01	1.11	0.52	2.64
Transport	*	208.12	*	208.12
Res./com.	0	0.90	0.63	1.53
Solvent evap.	*	*	*	375.00
Total	10.39	392.56	97.49	875.44

AUSTRIA

1977

VOC Emissions　　　　　　　　　　　　　　　　1000 Metric Tonnes

	Coal	Oil	Gas	Total
Production	0.03	1.80	7.29	9.12
Pet. ref.	*	5.32	*	5.32
Gasoline dist.	*	10.18	*	10.18
Electricity	0.11	0.11	0.02	0.24
Industry	0.09	1.89	0.13	2.11
Transport	*	121.98	*	121.98
Res./com.	9.92	0.50	0.11	10.53
Solvent evap.	*	*	*	75.00
Total	10.15	141.78	7.55	234.48

1985

VOC Emissions　　　　　　　　　　　　　　　　1000 Metric Tonnes

	Coal	Oil	Gas	Total
Production	0.01	1.70	3.60	5.31
Pet. ref.	*	5.25	*	5.25
Gasoline dist.	*	22.50	*	22.50
Electricity	0.08	0.21	0.03	0.32
Industry	0.14	2.16	0.11	2.41
Transport	*	141.15	*	141.15
Res./com.	6.40	0.66	0.29	7.35
Solvent evap.	*	*	*	104.00
Total	6.63	173.63	4.03	288.29

1990

VOC Emissions　　　　　　　　　　　　　　　　1000 Metric Tonnes

	Coal	Oil	Gas	Total
Production	0.01	1.60	3.24	4.85
Pet. ref.	*	5.82	*	5.82
Gasoline dist.	*	31.04	*	31.04
Electricity	0.11	0.48	0.04	0.63
Industry	0.14	2.62	0.11	2.87
Transport	*	132.66	*	132.66
Res./com.	5.60	0.80	0.36	6.76
Solvent evap.	*	*	*	124.00
Total	5.86	175.02	3.75	308.62

BELGIUM

1977

VOC Emissions   1000 Metric Tonnes

	Coal	Oil	Gas	Total
Production	0.14	0	0.11	0.25
Pet. ref.	*	10.98	*	10.98
Gasoline dist.	*	21.03	*	21.03
Electricity	0.72	0.43	0.03	1.18
Industry	0.36	48.81	0.28	49.45
Transport	*	192.75	*	192.75
Res./com.	10.49	1.06	0.40	11.95
Solvent evap.	*	*	*	127.00
Total	11.71	275.06	0.82	414.59

1985

VOC Emissions   1000 Metric Tonnes

	Coal	Oil	Gas	Total
Production	0.14	0	0	0.14
Pet. ref.	*	9.91	*	9.91
Gasoline dist.	*	42.45	*	42.45
Electricity	0.92	0.50	0.04	1.46
Industry	0.14	39.41	0.32	39.87
Transport	*	189.21	*	189.21
Res./com.	4.00	1.12	0.66	5.78
Solvent evap.	*	*	*	178.00
Total	5.20	282.60	1.02	466.82

1990

VOC Emissions   1000 Metric Tonnes

	Coal	Oil	Gas	Total
Production	0.14	0	0	0.14
Pet. ref.	*	10.83	*	10.83
Gasoline dist.	*	57.76	*	57.76
Electricity	1.18	0.74	0.02	1.94
Industry	0.14	55.58	0.33	56.05
Transport	*	193.03	*	193.03
Res./com.	4.00	1.20	0.75	5.95
Solvent evap.	*	*	*	220.00
Total	5.46	319.14	1.10	545.70

CANADA

1977

VOC Emissions　　　　　　　　　　　　　　　　　　1000 Metric Tonnes

	Coal	Oil	Gas	Total
Production	0.43	123.03	261.43	384.89
Pet. ref.	*	221.34	*	221.34
Gasoline dist.	*	265.61	*	265.61
Electricity	2.47	0.58	0.06	3.11
Industry	0.66	92.45	0.87	93.98
Transport	*	1035.73	*	1035.73
Res./com.	0.99	2.66	2.21	5.86
Solvent evap.	*	*	*	242.00
Total	4.55	1741.40	264.57	2252.52

1985

VOC Emissions　　　　　　　　　　　　　　　　　　1000 Metric Tonnes

	Coal	Oil	Gas	Total
Production	0.64	121.23	282.24	404.11
Pet. ref.	*	134.93	*	134.93
Gasoline dist.	*	309.00	*	309.00
Electricity	3.49	0.94	0.11	4.54
Industry	0.32	117.24	1.09	118.65
Transport	*	672.95	*	672.95
Res./com.	0	2.94	2.93	5.87
Solvent evap.	*	*	*	357.00
Total	4.45	1359.23	286.37	2007.05

1990

VOC Emissions　　　　　　　　　　　　　　　　　　1000 Metric Tonnes

	Coal	Oil	Gas	Total
Production	0.84	142.89	260.75	404.48
Pet. ref.	*	132.12	*	132.12
Gasoline dist.	*	352.32	*	352.32
Electricity	4.56	0.87	0.11	5.54
Industry	0.28	126.50	1.30	128.08
Transport	*	584.96	*	584.96
Res./com.	0	2.88	3.46	6.34
Solvent evap.	*	*	*	443.00
Total	5.68	1342.54	265.62	2056.84

DENMARK

1977

VOC Emissions   1000 Metric Tonnes

	Coal	Oil	Gas	Total
Production	0.01	0.51	0	0.52
Pet. ref.	*	7.71	*	7.71
Gasoline dist.	*	14.76	*	14.76
Electricity	0.61	0.37	0	0.98
Industry	0.23	6.69	0	6.92
Transport	*	116.57	*	116.57
Res./com.	1.81	1.00	0.01	2.82
Solvent evap.	*	*	*	72.00
Total	2.66	147.61	0.01	222.28

1985

VOC Emissions   1000 Metric Tonnes

	Coal	Oil	Gas	Total
Production	0	3.50	7.20	10.70
Pet. ref.	*	4.52	*	4.52
Gasoline dist.	*	19.35	*	19.35
Electricity	0.88	0.25	0	1.13
Industry	0.32	6.15	0.01	6.48
Transport	*	110.39	*	110.39
Res./com.	3.20	0.53	0.25	3.98
Solvent evap.	*	*	*	104.00
Total	4.40	144.69	7.46	260.55

1990

Voc Emissions   1000 Metric Tonnes

	Coal	Oil	Gas	Total
Production	0	1.90	9.72	11.62
Pet. ref.	*	3.60	*	3.60
Gasoline dist.	*	19.20	*	19.20
Electricity	1.34	0.10	0	1.44
Industry	0.32	6.74	0.02	7.09
Transport	*	106.81	*	106.81
Res./com.	3.20	0.46	0.33	3.99
Solvent evap.	*	*	*	129.00
Total	4.86	138.81	10.07	282.75

FINLAND

1977

VOC Emissions   1000 Metric Tonnes

	Coal	Oil	Gas	Total
Production	0.09	0	0	0.09
Pet. ref.	*	6.11	*	6.11
Gasoline dist.	*	11.70	*	11.70
Electricity	0.43	0.13	0.01	0.57
Industry	0.69	20.43	0.03	21.15
Transport	*	90.84	*	90.84
Res./com.	16.44	0.71	0	17.16
Solvent evap.	*	*	*	51.00
Total	17.65	129.92	0.04	198.62

1985

VOC Emissions   1000 Metric Tonnes

	Coal	Oil	Gas	Total
Production	0	0	0	0
Pet. ref.	*	5.29	*	5.29
Gasoline dist.	*	22.65	*	22.65
Electricity	0.25	0.21	0	0.46
Industry	0.60	33.65	0.04	34.29
Transport	*	103.29	*	103.29
Res./com.	3.20	0.59	0.06	3.85
Solvent evap.	*	*	*	74.00
Total	4.05	165.68	0.10	243.83

1990

VOC Emissions   1000 Metric Tonnes

	Coal	Oil	Gas	Total
Production	0	0	0	0
Pet. ref.	*	5.31	*	5.31
Gasoline dist.	*	28.32	*	28.32
Electricity	0.25	0.34	0	0.59
Industry	0.64	37.31	0.04	37.99
Transport	*	103.30	*	103.30
Res./com.	4.80	0.63	0.06	5.49
Solvent evap.	*	*	*	86.00
Total	5.69	175.21	0.10	267.00

FRANCE

1977

VOC Emissions   1000 Metric Tonnes

	Coal	Oil	Gas	Total
Production	0.46	1.04	23.61	25.11
Pet. ref.	*	49.79	*	49.79
Gasoline dist.	*	95.35	*	95.35
Electricity	3.36	1.28	0.04	4.68
Industry	0.91	206.77	0.49	208.17
Transport	*	1044.80	*	1044.80
Res./com.	31.94	4.28	1.14	37.36
Solvent evap.	*	*	*	679.00
Total	36.67	1403.31	25.28	2144.26

1985

VOC Emissions   1000 Metric Tonnes

	Coal	Oil	Gas	Total
Production	0.36	2.00	23.40	25.76
Pet. ref.	*	43.51	*	43.51
Gasoline dist.	*	186.45	*	186.45
Electricity	2.33	1.02	0.02	3.37
Industry	2.99	220.60	0.96	224.55
Transport	*	894.72	*	894.72
Res./com.	25.60	4.30	2.60	32.50
Solvent evap.	*	*	*	1000.00
Total	31.28	1352.60	26.98	2410.86

1990

VOC Emissions   1000 Metric Tonnes

	Coal	Oil	Gas	Total
Production	0.36	2.00	9.36	11.72
Pet. ref.	*	41.34	*	41.34
Gasoline dist.	*	220.48	*	220.48
Electricity	2.54	0.94	0.02	3.50
Industry	4.42	258.40	1.00	263.82
Transport	*	853.52	*	853.52
Res./com.	32.00	4.55	2.97	39.52
Solvent evap.	*	*	*	1237.00
Total	39.32	1381.23	13.35	2670.90

GERMANY

1977

VOC Emissions   1000 Metric Tonnes

	Coal	Oil	Gas	Total
Production	2.34	5.45	53.69	61.48
Pet. ref.	*	65.21	*	65.21
Gasoline dist.	*	124.87	*	124.87
Electricity	9.88	0.87	0.27	11.02
Industry	1.90	268.03	0.82	270.75
Transport	*	1206.14	*	1206.14
Res./com.	44.27	6.56	1.94	52.77
Solvent evap.	*	*	*	860.00
Total	58.39	1677.13	56.72	2652.24

1985

VOC Emissions   1000 Metric Tonnes

	Coal	Oil	Gas	Total
Production	2.39	3.80	52.92	59.11
Pet. ref.	*	56.32	*	56.32
Gasoline dist.	*	241.35	*	241.35
Electricity	11.19	2.06	0.34	13.59
Industry	1.89	355.02	1.56	358.47
Transport	*	1034.52	*	1034.52
Res./com.	16.00	5.46	3.22	24.68
Solvent evap.	*	*	*	1185.00
Total	31.47	1698.53	58.04	2973.04

1990

VOC Emissions   1000 Metric Tonnes

	Coal	Oil	Gas	Total
Production	2.37	3.50	52.92	58.79
Pet. ref.	*	51.39	*	51.39
Gasoline dist.	*	274.08	*	274.08
Electricity	13.44	1.97	0.34	15.75
Industry	1.79	381.05	1.60	384.44
Transport	*	963.65	*	963.65
Res./com.	16.00	5.68	3.43	25.11
Solvent evap.	*	*	*	1374.00
Total	33.60	1681.32	58.29	3147.21

GREECE

1977

VOC Emissions   1000 Metric Tonnes

	Coal	Oil	Gas	Total
Production	0.09	0	0	0.09
Pet. ref.	*	4.76	*	4.76
Gasoline dist.	*	9.11	*	9.11
Electricity	0.63	0.19	0	0.82
Industry	0.09	5.50	0	5.59
Transport	*	104.70	*	104.70
Res./com.	0.38	0.25	0	0.63
Solvent evap.	*	*	*	43.00
Total	1.19	124.51	0	168.70

1985

VOC Emissions   1000 Metric Tonnes

	Coal	Oil	Gas	Total
Production	0.20	1.00	1.44	2.64
Pet. ref.	*	4.90	*	4.90
Gasoline dist.	*	21.00	*	21.00
Electricity	1.43	0.22	0	1.65
Industry	0.09	8.96	0	9.05
Transport	*	104.13	*	104.13
Res./com.	0	0.45	0.02	0.47
Solvent evap.	*	*	*	74.00
Total	1.72	140.66	1.46	217.84

1990

VOC Emissions   1000 Metric Tonnes

	Coal	Oil	Gas	Total
Production	0.24	1.00	0.72	1.96
Pet. ref.	*	5.70	*	5.70
Gasoline dist.	*	30.40	*	30.40
Electricity	1.74	0.50	0	2.24
Industry	0.09	11.73	0	11.82
Transport	*	104.76	*	104.76
Res./com.	0	0.53	0.02	0.55
Solvent evap.	*	*	*	96.00
Total	2.08	154.62	0.74	253.43

ICELAND

1977

VOC Emissions   1000 Metric Tonnes

	Coal	Oil	Gas	Total
Production	0	0	C	0
Pet. ref.	*	0.31	*	0.31
Gasoline dist.	*	0.58	*	0.58
Electricity	0	0.01	0	0.01
Industry	0	0.01	0	0.01
Transport	*	10.83	*	10.83
Res./com.	0	0.02	0	0.02
Solvent evap.	*	*	*	2.00
Total	0	11.76	0	13.76

1985

VOC Emissions   1000 Metric Tonnes

	Coal	Oil	Gas	Total
Production	0	0	0	0
Pet. ref.	*	0.28	*	0.28
Gasoline dist.	*	1.20	*	1.20
Electricity	0	0.01	0	0.01
Industry	0	0.01	0	0.01
Transport	*	10.45	*	10.45
Res./com.	0	0.03	0	0.03
Solvent evap.	*	*	*	3.00
Total	0	11.98	0	14.98

1990

VOC Emissions   1000 Metric Tonnes

	Coal	Oil	Gas	Total
Production	0	0	0	0
Pet. ref.	*	0.33	*	0.33
Gasoline dist.	*	1.76	*	1.76
Electricity	0	0.01	0	0.01
Industry	0	0.03	0	0.03
Transport	*	11.72	*	11.72
Res./com.	0	0.04	0	0.04
Solvent evap.	*	*	*	4.00
Total	0	13.89	0	17.89

IRELAND

1977

VOC Emissions 1000 Metric Tonnes

	Coal	Oil	Gas	Total
Production	0.03	0	0	0.03
Pet. ref.	*	2.57	*	2.57
Gasoline dist.	*	4.92	*	4.92
Electricity	0.14	0.19	0	0.33
Industry	0	0.21	0	0.21
Transport	*	53.52	*	53.52
Res./com.	9.39	0.11	0.01	9.51
Solvent evap.	*	*	*	16.00
Total	9.56	61.52	0.01	87.09

1985

VOC Emissions 1000 Metric Tonnes

	Coal	Oil	Gas	Total
Production	0	0	3.60	3.60
Pet. ref.	*	3.29	*	3.29
Gasoline dist.	*	14.10	*	14.10
Electricity	0	0.36	0.01	0.38
Industry	0	0.53	0	0.53
Transport	*	54.93	*	54.93
Res./com.	4.80	0.11	0	4.91
Solvent evap.	*	*	*	22.00
Total	4.80	73.32	3.61	103.74

1990

VOC Emissions 1000 Metric Tonnes

	Coal	Oil	Gas	Total
Production	0	0	3.60	3.60
Pet. ref.	*	4.77	*	4.77
Gasoline dist.	*	25.44	*	25.44
Electricity	0	0.67	0.01	0.68
Industry	0	0.94	0	0.94
Transport	*	64.61	*	64.61
Res./com.	4.80	0.20	0	5.00
Solvent evap.	*	*	*	27.00
Total	4.80	96.63	3.61	132.04

ITALY

1977

VOC Emissions 1000 Metric Tonnes

	Coal	Oil	Gas	Total
Production	0.03	1.13	41.51	42.67
Pet. ref.	*	43.25	*	43.25
Gasoline dist.	*	82.82	*	82.82
Electricity	0.53	2.76	0.06	3.35
Industry	0.34	262.29	0.69	263.32
Transport	*	709.40	*	709.40
Res./com.	8.92	2.80	1.07	12.79
Solvent evap.	*	*	*	350.00
Total	9.82	1104.45	43.33	1507.60

1985

VOC Emissions 1000 Metric Tonnes

	Coal	Oil	Gas	Total
Production	0.04	3.00	54.00	57.04
Pet. ref.	*	44.38	*	44.38
Gasoline dist.	*	190.20	*	190.20
Electricity	1.22	5.31	0.04	6.57
Industry	0.09	223.50	1.03	224.62
Transport	*	710.18	*	710.18
Res./com.	7.20	3.92	1.72	12.84
Solvent evap.	*	*	*	492.00
Total	8.55	1180.49	56.79	1737.83

1990

VOC Emissions 1000 Metric Tonnes

	Coal	Oil	Gas	Total
Production	0.04	2.40	48.24	50.68
Pet. ref.	*	43.77	*	43.77
Gasoline dist.	*	233.44	*	233.44
Electricity	1.89	6.79	0.01	8.69
Industry	0.14	248.57	1.21	249.92
Transport	*	644.73	*	644.73
Res./com.	5.60	4.12	1.97	11.69
Solvent evap.	*	*	*	584.00
Total	7.67	1183.82	51.43	1826.92

JAPAN

1977

VOC Emissions   1000 Metric Tonnes

	Coal	Oil	Gas	Total
Production	0.40	0.60	9.07	10.07
Pet. ref.	*	120.12	*	120.12
Gasoline dist.	*	230.02	*	230.02
Electricity	2.66	9.09	0.13	11.88
Industry	1.03	1011.72	0.16	1012.91
Transport	*	1132.62	*	1132.62
Res./com.	26.87	5.49	0.94	33.30
Solvent evap.	*	*	*	1139.00
Total	30.96	2509.66	10.30	3689.92

1985

VOC Emissions   1000 Metric Tonnes

	Coal	Oil	Gas	Total
Production	0.37	4.20	22.32	26.89
Pet. ref.	*	140.95	*	140.95
Gasoline dist.	*	604.05	*	604.05
Electricity	4.33	15.85	0.66	20.84
Industry	1.29	1891.99	0.13	1893.41
Transport	*	880.40	*	880.40
Res./com.	0	8.05	1.44	9.49
Solvent evap.	*	*	*	1898.00
Total	5.99	3545.49	24.55	5474.03

1990

VOC Emissions   1000 Metric Tonnes

	Coal	Oil	Gas	Total
Production	0.37	5.20	28.08	33.65
Pet. ref.	*	131.97	*	131.97
Gasoline dist.	*	703.84	*	703.84
Electricity	7.18	16.02	0.89	24.09
Industry	1.56	2265.63	0.19	2267.38
Transport	*	769.93	*	769.93
Res./com.	0	8.20	2.16	10.36
Solvent evap.	*	*	*	2422.00
Total	9.11	3900.79	31.32	6363.22

LUXEMBURG

1977

VOC Emissions   1000 Metric Tonnes

	Coal	Oil	Gas	Total
Production	0	0	0	0
Pet. ref.	*	0.68	*	0.68
Gasoline dist.	*	1.30	*	1.30
Electricity	0.02	0.01	0	0.03
Industry	0.01	0.49	0.01	0.51
Transport	*	13.02	*	13.02
Res./com.	0.19	0.06	0.01	0.26
Solvent evap.	*	*	*	4.00
Total	0.22	15.55	0.02	19.80

1985

VOC Emissions   1000 Metric Tonnes

	Coal	Oil	Gas	Total
Production	0	0	0	0
Pet. ref.	*	1.12	*	1.12
Gasoline dist.	*	4.80	*	4.80
Electricity	0.06	0.14	0	0.20
Industry	0.32	1.12	0.01	1.45
Transport	*	13.85	*	13.85
Res./com.	0	0.07	0.05	0.12
Solvent evap.	*	*	*	6.00
Total	0.38	21.10	0.06	27.54

1990

VOC Emissions   1000 Metric Tonnes

	Coal	Oil	Gas	Total
Production	0	0	0	0
Pet. ref.	*	1.29	*	1.29
Gasoline dist.	*	6.88	*	6.88
Electricity	0.06	0.31	0	0.37
Industry	0.37	0.94	0.01	1.32
Transport	*	13.81	*	13.81
Res./com.	0	0.07	0.05	0.12
Solvent evap.	*	*	*	7.00
Total	0.43	23.30	0.06	30.79

NETHERLANDS

1977

VOC Emissions   1000 Metric Tonnes

	Coal	Oil	Gas	Total
Production	0	1.64	268.70	270.34
Pet. ref.	*	12.48	*	12.48
Gasoline dist.	*	23.89	*	23.89
Electricity	0.29	0.11	0.20	0.60
Industry	0.10	238.34	0.56	239.00
Transport	*	286.32	*	286.32
Res./com.	0.63	0.62	2.13	3.38
Solvent evap.	*	*	*	163.00
Total	1.02	563.40	271.59	999.01

1985

VOC Emissions   1000 Metric Tonnes

	Coal	Oil	Gas	Total
Production	0	1.50	249.48	250.98
Pet. ref.	*	16.84	*	16.84
Gasoline dist.	*	72.15	*	72.15
Electricity	0.69	0.94	0.11	1.74
Industry	0.05	579.68	0.33	580.06
Transport	*	286.49	*	286.49
Res./com.	0	0.63	3.04	3.67
Solvent evap.	*	*	*	224.00
Total	0.74	958.23	252.96	1435.93

1990

VOC Emissions   1000 Metric Tonnes

	Coal	Oil	Gas	Total
Production	0	1.60	165.60	167.20
Pet. ref.	*	17.58	*	17.58
Gasoline dist.	*	93.76	*	93.76
Electricity	1.85	1.78	0.06	3.69
Industry	0.51	696.81	0.30	697.62
Transport	*	284.43	*	284.43
Res./com.	0	0.83	3.06	3.89
Solvent evap.	*	*	*	257.00
Total	2.36	1096.79	169.02	1525.17

NEW ZEALAND

1977

VOC Emissions   1000 Metric Tonnes

	Coal	Oil	Gas	Total
Production	0.06	0.68	5.08	5.82
Pet. ref.	*	2.01	*	2.01
Gasoline dist.	*	3.85	*	3.85
Electricity	0.09	0.03	0.02	0.14
Industry	0.50	0.08	0.01	0.59
Transport	*	87.44	*	87.44
Res./com.	2.91	0.07	0.01	2.99
Solvent evap.	*	*	*	26.00
Total	3.56	94.16	5.12	128.84

1985

VOC Emissions   1000 Metric Tonnes

	Coal	Oil	Gas	Total
Production	0.04	0.60	10.80	11.44
Pet. ref.	*	1.40	*	1.40
Gasoline dist.	*	6.00	*	6.00
Electricity	0.06	0.01	0.03	0.10
Industry	0.46	0.07	0.04	0.57
Transport	*	78.16	*	78.16
Res./com.	2.40	0.06	0.05	2.51
Solvent evap.	*	*	*	36.00
Total	2.96	86.30	10.92	136.18

1990

VOC Emissions   1000 Metric Tonnes

	Coal	Oil	Gas	Total
Production	0.07	0.70	14.40	15.17
Pet. ref.	*	1.32	*	1.32
Gasoline dist.	*	7.04	*	7.04
Electricity	0.21	0.01	0.05	0.27
Industry	0.60	0.08	0.06	0.74
Transport	*	71.16	*	71.16
Res./com.	2.40	0.03	0.06	2.49
Solvent evap.	*	*	*	42.00
Total	3.28	80.34	14.57	140.19

NORWAY

1977

VOC Emissions 1000 Metric Tonnes

	Coal	Oil	Gas	Total
Production	0.01	13.78	9.70	23.49
Pet. ref.	*	3.79	*	3.79
Gasoline dist.	*	7.26	*	7.26
Electricity	0	0	0	0
Industry	0.07	11.83	0	11.90
Transport	*	84.99	*	84.99
Res./com.	0.60	0.27	0	0.87
Solvent evap.	*	*	*	58.00
Total	0.68	121.92	9.70	190.30

1985

VOC Emissions 1000 Metric Tonnes

	Coal	Oil	Gas	Total
Production	0.03	39.00	97.20	136.23
Pet. ref.	*	4.06	*	4.06
Gasoline dist.	*	17.40	*	17.40
Electricity	0	0	0	0
Industry	0.32	13.78	0	14.10
Transport	*	98.39	*	98.39
Res./com.	0.80	0.50	0	1.30
Solvent evap.	*	*	*	88.00
Total	1.15	173.13	97.20	359.48

1990

VOC Emissions 1000 Metric Tonnes

	Coal	Oil	Gas	Total
Production	0.03	48.50	100.08	148.61
Pet. ref.	*	3.93	*	3.93
Gasoline dist.	*	20.96	*	20.96
Electricity	0	0	0	0
Industry	0.37	15.26	0	15.63
Transport	*	101.69	*	101.69
Res./com.	0.80	0.55	0.11	1.46
Solvent evap.	*	*	*	108.00
Total	1.20	190.89	100.19	400.28

PORTUGAL

1977

VOC Emissions　　　　　　　　　　　　　　　　　　　　1000 Metric Tonnes

	Coal	Oil	Gas	Total
Production	0	0.09	0	0.09
Pet. ref.	*	3.04	*	3.04
Gasoline dist.	*	5.81	*	5.81
Electricity	0.02	0.10	0	0.12
Industry	0.02	11.53	0	11.55
Transport	*	78.38	*	78.38
Res./com.	0.08	0.10	0.01	0.19
Solvent evap.	*	*	*	30.00
Total	0.12	99.05	0.01	129.18

1985

VOC Emissions　　　　　　　　　　　　　　　　　　　　1000 Metric Tonnes

	Coal	Oil	Gas	Total
Production	0.01	0	0	0.01
Pet. ref.	*	3.75	*	3.75
Gasoline dist.	*	16.05	*	16.05
Electricity	0.19	0.17	0	0.36
Industry	0	22.64	0	22.64
Transport	*	76.65	*	76.65
Res./com.	0	0.21	0	0.21
Solvent evap.	*	*	*	44.00
Total	0.20	119.47	0	163.67

1990

VOC Emissions　　　　　　　　　　　　　　　　　　　　1000 Metric Tonnes

	Coal	Oil	Gas	Total
Production	0.01	0	0	0.01
Pet. ref.	*	4.26	*	4.26
Gasoline dist.	*	22.72	*	22.72
Electricity	0.31	0.38	0	0.69
Industry	0	25.70	0	25.70
Transport	*	78.54	*	78.54
Res./com.	0	0.27	0	0.27
Solvent evap.	*	*	*	56.00
Total	0.32	131.87	0	188.19

SPAIN

1977

VOC Emissions   1000 Metric Tonnes

	Coal	Oil	Gas	Total
Production	0.24	0.84	0	1.09
Pet. ref.	*	20.72	*	20.72
Gasoline dist.	*	39.67	*	39.67
Electricity	1.24	0.84	0	2.08
Industry	0.32	80.53	0.04	80.89
Transport	*	462.68	*	462.68
Res./com.	2.48	1.02	0.06	3.56
Solvent evap.	*	*	*	202.00
Total	4.28	606.30	0.10	812.69

1985

VOC Emissions   1000 Metric Tonnes

	Coal	Oil	Gas	Total
Production	0.36	7.60	0	7.96
Pet. ref.	*	19.11	*	19.11
Gasoline dist.	*	81.90	*	81.90
Electricity	1.79	1.29	0	3.08
Industry	0.51	77.23	0.24	77.98
Transport	*	433.24	*	433.24
Res./com.	3.20	1.05	0.36	4.61
Solvent evap.	*	*	*	265.00
Total	5.86	621.42	0.60	892.88

1990

VOC Emissions   1000 Metric Tonnes

	Coal	Oil	Gas	Total
Production	0.43	9.00	0	9.43
Pet. ref.	*	19.77	*	19.77
Gasoline dist.	*	105.44	*	105.44
Electricity	2.25	1.26	0	3.51
Industry	0.51	98.39	0.34	99.24
Transport	*	427.13	*	427.13
Res./com.	3.20	1.36	0.66	5.22
Solvent evap.	*	*	*	328.00
Total	6.39	662.35	1.00	997.74

SWEDEN

1977

VOC Emissions   1000 Metric Tonnes

	Coal	Oil	Gas	Total
Production	0.08	0	0	0.08
Pet. ref.	*	12.87	*	12.87
Gasoline dist.	*	24.65	*	24.65
Electricity	0.04	0.36	0	0.40
Industry	1.43	27.84	0	29.27
Transport	*	209.06	*	209.06
Res./com.	0.12	1.55	0.01	1.68
Solvent evap.	*	*	*	101.00
Total	1.67	276.33	0.01	379.01

1985

VOC Emissions   1000 Metric Tonnes

	Coal	Oil	Gas	Total
Production	0	0	0	0
Pet. ref.	*	11.48	*	11.48
Gasoline dist.	*	49.20	*	49.20
Electricity"	0	0.87	0	0.87
Industry	0.78	31.67	0	32.45
Transport	*	176.71	*	176.71
Res./com.	0	1.37	0.02	1.39
Solvent·evap.	*	*	*	141.00
Total	0.78	271.30	0.02	413.10

1990

VOC Emissions   1000 Metric Tonnes

	Coal	Oil	Gas	Total
Production	0	0	0	0
Pet. ref.	*	9.15	*	9.15
Gasoline dist.	*	48.80	*	48.80
Electricity	0	0.59	0	0.59
Industry	0.83	32.12	0	32.95
Transport	*	151.27	*	151.27
Res./com.	0	1.23	0.02	1.25
Solvent evap.	*	*	*	161.00
Total	0.83	243.16	0.02	405.01

SWITZERLAND

1977

VOC Emissions   1000 Metric Tonnes

	Coal	Oil	Gas	Total
Production	0.01	0	0	0.01
Pet. ref.	*	6.34	*	6.34
Gasoline dist.	*	12.15	*	12.15
Electricity	0	0.06	0	0.06
Industry	0.04	4.18	0.02	4.24
Transport	*	138.02	*	138.02
Res./com.	2.41	0.85	0.04	3.30
Solvent evap.	*	*	*	102.00
Total	2.46	161.60	0.06	266.12

1985

VOC Emissions   1000 Metric Tonnes

	Coal	Oil	Gas	Total
Production	0	0	0	0
Pet. ref.	*	5.39	*	5.39
Gasoline dist.	*	23.10	*	23.10
Electricity	0.02	0.34	0	0.36
Industry	0.09	3.82	0.04	3.95
Transport	*	129.72	*	129.72
Res./com.	0.80	0.74	0.12	1.66
Solvent evap.	*	*	*	134.00
Total	0.91	163.11	0.16	298.18

1990

VOC Emissions   1000 Metric Tonnes

	Coal	Oil	Gas	Total
Production	0	0	0	0
Pet. ref.	*	4.62	*	4.62
Gasoline dist.	*	24.64	*	24.64
Electricity	0.06	0.15	0	0.21
Industry	0.14	4.51	0.04	4.69
Transport	*	117.50	*	117.50
Res./com.	0	0.81	0.15	0.96
Solvent evap.	*	*	*	154.00
Total	0.20	152.23	0.19	306.62

TURKEY

1977

VOC Emissions 1000 Metric Tonnes

	Coal	Oil	Gas	Total
Production	0.35	2.62	0	2.97
Pet. ref.	*	8.09	*	8.09
Gasoline dist.	*	15.49	*	15.49
Electricity	0.34	0.26	0	0.60
Industry	0.28	15.37	0	15.65
Transport	*	219.05	*	219.05
Res./com.	70.46	0.58	0	71.04
Solvent evap.	*	*	*	76.00
Total	71.43	261.46	0	408.89

1985

VOC Emissions 1000 Metric Tonnes

	Coal	Oil	Gas	Total
Production	0.54	3.20	0	3.74
Pet. ref.	*	10.57	*	10.57
Gasoline dist.	*	45.30	*	45.30
Electricity	1.79	0.18	0	1.97
Industry	1.89	23.08	0	24.97
Transport	*	248.81	*	248.81
Res./com.	56.00	1.39	0	57.39
Solvent evap.	*	*	*	147.00
Total	60.22	332.53	0	539.75

1990

VOC Emissions 1000 Metric Tonnes

	Coal	Oil	Gas	Total
Production	0.65	3.10	0	3.75
Pet. ref.	*	12.60	*	12.60
Gasoline dist.	*	67.20	*	67.20
Electricity	2.58	0.15	0	2.73
Industry	2.35	37.10	0	39.45
Transport	*	290.15	*	290.15
Res./com.	76.80	1.78	0	78.58
Solvent evap.	*	*	*	206.00
Total	82.38	412.08	0	700.46

UNITED KINGDOM

1977

VOC Emissions   1000 Metric Tonnes

	Coal	Oil	Gas	Total
Production	1.93	39.19	124.98	166.10
Pet. ref.	*	43.52	*	43.52
Gasoline dist.	*	83.33	*	83.33
Electricity	9.51	1.76	0.03	11.30
Industry	2.08	211.47	0.81	214.36
Transport	*	1056.65	*	1056.65
Res./com.	90.15	1.72	2.83	94.70
Solvent evap.	*	*	*	319.00
Total	103.67	1437.64	128.65	1988.96

1985

VOC Emissions   1000 Metric Tonnes

	Coal	Oil	Gas	Total
Production	2.14	150.00	144.00	296.14
Pet. ref.	*	32.80	*	32.80
Gasoline dist.	*	140.55	*	140.55
Electricity	10.50	1.41	0.04	11.95
Industry	1.98	171.06	1.02	174.06
Transport	*	978.20	*	978.20
Res./com.	32.00	1.40	3.75	37.15
Solvent evap.	*	*	*	437.00
Total	46.62	1475.42	148.81	2107.85

1990

VOC Emissions   1000 Metric Tonnes

	Coal	Oil	Gas	Total
Production	2.14	140.00	162.00	304.14
Pet. ref.	*	29.25	*	29.25
Gasoline dist.	*	156.00	*	156.00
Electricity	12.18	0.38	0.04	12.60
Industry	0.46	185.53	1.11	187.10
Transport	*	899.03	*	899.03
Res./com.	8.00	1.68	4.13	13.81
Solvent evap.	*	*	*	509.00
Total	22.78	1411.87	167.28	2110.93

UNITED STATES

1977

VOC Emissions   1000 Metric Tonnes

	Coal	Oil	Gas	Total
Production	10.94	752.36	1743.75	2507.05
Pet. ref.	*	2197.71	*	2197.71
Gasoline dist.	*	2637.26	*	2637.26
Electricity	56.67	13.30	1.50	71.47
Industry	17.48	2111.89	8.36	2137.73
Transport	*	11935.30	*	11935.30
Res./com.	32.76	21.92	26.15	80.83
Solvent evap.	*	*	*	5743.00
Total	117.85	19669.74	1779.76	27310.35

1985

VOC Emissions   1000 Metric Tonnes

	Coal	Oil	Gas	Total
Production	16.18	822.15	1492.80	2331.13
Pet. ref.	*	1363.84	*	1363.84
Gasoline dist.	*	3123.30	*	3123.30
Electricity	90.49	15.71	0.80	107.00
Industry	19.37	3829.13	10.77	3859.27
Transport	*	7274.91	*	7274.91
Res./com.	0	19.53	26.47	46.00
Solvent evap.	*	*	*	9507.00
Total	126.04	16448.57	1530.84	27612.45

1990

VOC Emissions   1000 Metric Tonnes

	Coal	Oil	Gas	Total
Production	19.56	728.00	1502.38	2249.94
Pet. ref.	*	1358.76	*	1358.76
Gasoline dist.	*	3623.36	*	3623.36
Electricity	109.62	14.39	0.76	124.77
Industry	24.01	4834.93	12.09	4871.03
Transport	*	6097.19	*	6097.19
Res./com.	0	20.08	25.30	45.38
Solvent evap.	*	*	*	10968.00
Total	153.19	16676.71	1540.53	29338.43

OECD SALES AGENTS
DÉPOSITAIRES DES PUBLICATIONS DE L'OCDE

ARGENTINA – ARGENTINE
Carlos Hirsch S.R.L., Florida 165, 4° Piso (Galería Guemes)
1333 BUENOS AIRES, Tel. 33.1787.2391 y 30.7122

AUSTRALIA – AUSTRALIE
Australia and New Zealand Book Company Pty, Ltd.,
10 Aquatic Drive, Frenchs Forest, N.S.W. 2086
P.O. Box 459, BROOKVALE, N.S.W. 2100

AUSTRIA – AUTRICHE
OECD Publications and Information Center
4 Simrockstrasse 5300 BONN. Tel. (0228) 21.60.45
Local Agent/Agent local :
Gerold and Co., Graben 31, WIEN 1. Tel. 52.22.35

BELGIUM – BELGIQUE
LCLS
35, avenue de Stalingrad, 1000 BRUXELLES. Tel. 02.512.89.74

BRAZIL – BRÉSIL
Mestre Jou S.A., Rua Guaipa 518,
Caixa Postal 24090, 05089 SAO PAULO 10. Tel. 261.1920
Rua Senador Dantas 19 s/205-6, RIO DE JANEIRO GB.
Tel. 232.07.32

CANADA
Renouf Publishing Company Limited,
2182 St. Catherine Street West,
MONTRÉAL, Que. H3H 1M7. Tel. (514)937.3519
OTTAWA, Ont. K1P 5A6, 61 Sparks Street

DENMARK – DANEMARK
Munksgaard Export and Subscription Service
35, Nørre Søgade
DK 1370 KØBENHAVN K. Tel. +45.1.12.85.70

FINLAND – FINLANDE
Akateeminen Kirjakauppa
Keskuskatu 1, 00100 HELSINKI 10. Tel. 65.11.22

FRANCE
Bureau des Publications de l'OCDE,
2 rue André-Pascal, 75775 PARIS CEDEX 16. Tel. (1) 524.81.67
Principal correspondant :
13602 AIX-EN-PROVENCE : Librairie de l'Université.
Tel. 26.18.08

GERMANY – ALLEMAGNE
OECD Publications and Information Center
4 Simrockstrasse 5300 BONN Tel. (0228) 21.60.45

GREECE – GRÈCE
Librairie Kauffmann, 28 rue du Stade,
ATHÈNES 132. Tel. 322.21.60

HONG-KONG
Government Information Services,
Publications/Sales Section, Baskerville House,
2/F., 22 Ice House Street

ICELAND – ISLANDE
Snaebjörn Jönsson and Co., h.f.,
Hafnarstraeti 4 and 9, P.O.B. 1131, REYKJAVIK.
Tel. 13133/14281/11936

INDIA – INDE
Oxford Book and Stationery Co. :
NEW DELHI-1, Scindia House. Tel. 45896
CALCUTTA 700016, 17 Park Street. Tel. 240832

INDONESIA – INDONÉSIE
PDIN-LIPI, P.O. Box 3065/JKT., JAKARTA, Tel. 583467

IRELAND – IRLANDE
TDC Publishers – Library Suppliers
12 North Frederick Street, DUBLIN 1 Tel. 744835-749677

ITALY – ITALIE
Libreria Commissionaria Sansoni :
Via Lamarmora 45, 50121 FIRENZE. Tel. 579751
Via Bartolini 29, 20155 MILANO. Tel. 365083
Sub-depositari :
Editrice e Libreria Herder,
Piazza Montecitorio 120, 00 186 ROMA. Tel. 6794628
Libreria Hoepli, Via Hoepli 5, 20121 MILANO. Tel. 865446
Libreria Lattes, Via Garibaldi 3, 10122 TORINO. Tel. 519274
La diffusione delle edizioni OCSE è inoltre assicurata dalle migliori
librerie nelle città più importanti.

JAPAN – JAPON
OECD Publications and Information Center,
Landic Akasaka Bldg., 2-3-4 Akasaka,
Minato-ku, TOKYO 107 Tel. 586.2016

KOREA – CORÉE
Pan Korea Book Corporation,
P.O. Box n° 101 Kwangwhamun, SÉOUL. Tel. 72.7369

LEBANON – LIBAN
Documenta Scientifica/Redico,
Edison Building, Bliss Street, P.O. Box 5641, BEIRUT.
Tel. 354429 – 344425

MALAYSIA – MALAISIE
and/et SINGAPORE - SINGAPOUR
University of Malaysia Co-operative Bookshop Ltd.
P.O. Box 1127, Jalan Pantai Baru
KUALA LUMPUR. Tel. 51425, 54058, 54361

THE NETHERLANDS – PAYS-BAS
Staatsuitgeverij
Verzendboekhandel Chr. Plantijnstraat 1
Postbus 20014
2500 EA S-GRAVENAGE. Tel. nr. 070.789911
Voor bestellingen: Tel. 070.789208

NEW ZEALAND – NOUVELLE-ZÉLANDE
Publications Section,
Government Printing Office Bookshops:
AUCKLAND: Retail Bookshop: 25 Rutland Street,
Mail Orders: 85 Beach Road, Private Bag C.P.O.
HAMILTON: Retail Ward Street,
Mail Orders, P.O. Box 857
WELLINGTON: Retail: Mulgrave Street (Head Office),
Cubacade World Trade Centre
Mail Orders: Private Bag
CHRISTCHURCH: Retail: 159 Hereford Street,
Mail Orders: Private Bag
DUNEDIN: Retail: Princes Street
Mail Order: P.O. Box 1104

NORWAY – NORVÈGE
J.G. TANUM A/S Karl Johansgate 43
P.O. Box 1177 Sentrum OSLO 1. Tel. (02) 80.12.60

PAKISTAN
Mirza Book Agency, 65 Shahrah Quaid-E-Azam, LAHORE 3.
Tel. 66839

PHILIPPINES
National Book Store, Inc.
Library Services Division, P.O. Box 1934, MANILA.
Tel. Nos. 49.43.06 to 09, 40.53.45, 49.45.12

PORTUGAL
Livraria Portugal, Rua do Carmo 70-74,
1117 LISBOA CODEX. Tel. 360582/3

SPAIN – ESPAGNE
Mundi-Prensa Libros, S.A.
Castelló 37, Apartado 1223, MADRID-1. Tel. 275.46.55
Libreria Bastinos, Pelayo 52, BARCELONA 1. Tel. 222.06.00

SWEDEN – SUÈDE
AB CE Fritzes Kungl Hovbokhandel,
Box 16 356, S 103 27 STH, Regeringsgatan 12,
DS STOCKHOLM. Tel. 08/23.89.00

SWITZERLAND – SUISSE
OECD Publications and Information Center
4 Simrockstrasse 5300 BONN. Tel. (0228) 21.60.45
Local Agents/Agents locaux
Librairie Payot, 6 rue Grenus, 1211 GENÈVE 11. Tel. 022.31.89.50
Freihofer A.G., Weinbergstr. 109, CH-8006 ZÜRICH.
Tel. 01.3634282

THAILAND – THAILANDE
Suksit Siam Co., Ltd., 1715 Rama IV Rd,
Samyan, BANGKOK 5. Tel. 2511630

TURKEY – TURQUIE
Kültur Yayinlari Is-Türk Ltd. Sti.
Atatürk Bulvari No : 77/B
KIZILAY/ANKARA. Tel. 17 02 66
Dolmabahce Cad. No : 29
BESIKTAS/ISTANBUL. Tel. 60 71 88

UNITED KINGDOM – ROYAUME-UNI
H.M. Stationery Office, P.O.B. 569,
LONDON SE1   9NH. Tel. 01.928.6977, Ext. 410 or
49 High Holborn, LONDON WC1V 6 HB (personal callers)
Branches at: EDINBURGH, BIRMINGHAM, BRISTOL,
MANCHESTER, CARDIFF, BELFAST.

UNITED STATES OF AMERICA – ÉTATS-UNIS
OECD Publications and Information Center, Suite 1207,
1750 Pennsylvania Ave., N.W. WASHINGTON, D.C.20006 – 4582
Tel. (202) 724.1857

VENEZUELA
Libreria del Este, Avda. F. Miranda 52, Edificio Galipan,
CARACAS 106. Tel. 32.23.01/33.26.04/33.24.73

YUGOSLAVIA – YOUGOSLAVIE
Jugoslovenska Knjiga, Terazije 27, P.O.B. 36, BEOGRAD.
Tel. 621.992

Les commandes provenant de pays où l'OCDE n'a pas encore désigné de dépositaire peuvent être adressées à :
OCDE, Bureau des Publications, 2, rue André-Pascal, 75775 PARIS CEDEX 16.
Orders and inquiries from countries where sales agents have not yet been appointed may be sent to:
OECD, Publications Office, 2 rue André-Pascal, 75775 PARIS CEDEX 16.

OECD PUBLICATIONS, 2, rue André-Pascal, 75775 PARIS CEDEX 16 - No. 42135 1982
PRINTED IN FRANCE
(97 82 02 1) ISBN 92-64-12297-4